KU-795-596

HACCP

HACCP

A practical approach

Sara Mortimore

and

Carol Wallace

A Chapman & Hall Food Science Book

An Aspen Publication®
Aspen Publishers, Inc.
Gaithersburg, Maryland
1998

Copyright © 1998 by Sara Mortimore and Carol Wallace.
All rights reserved.

Aspen Publishers, Inc., grants permission for photocopying for limited personal or internal use. This consent does not extend to other kinds of copying, such as copying for general distribution, for advertising or promotional purposes, for creating new collective works, or for resale. For information, address Aspen Publishers, Inc., Permissions Department, 200 Orchard Ridge Drive, Suite 200, Gaithersburg, Maryland 20878.

Orders: (800) 638-8437
Customer Service: (800) 234-1660

About Aspen Publishers • For more than 40 years, Aspen has been a leading professional publisher in a variety of disciplines. Aspen's vast information resources are available in both print and electronic formats. We are committed to providing the highest quality information available in the most appropriate format for our customers. Visit Aspen's Internet site for more information resources, directories, articles, and a searchable version of Aspen's full catalog, including the most recent publications: **www.aspenpublishers.com**
Aspen Publishers, Inc. • The hallmark of quality in publishing
Member of the worldwide Wolters Kluwer group

Editorial Services: Rose Gulliver
Library of Congress Catalog Card Number: 98-70684
ISBN: 0-8342-1932-8

Printed in the United States of America
2 3 4 5

To Bill for continued inspiration, and to Lawrence and Christopher for their unfailing support and encouragement.

Contents

Contents

Authors' introduction to the second edition

When we wrote the first edition of *HACCP: A practical approach*, we did not dream that it would be so successful. Its popularity related mainly to the easy-to-read style, and step-by-step approach to planning, developing, implementing and maintaining an effective HACCP system.

There have been a number of changes in the HACCP field in the past four years, which this new book takes into account. The Codex *HACCP system and guidelines*, the international HACCP 'standard', has been updated and republished (Codex, 1997), and increased experience in the practicalities of HACCP worldwide has led to changes in the way it is applied. Specifically, this has led to the use of more modular HACCP systems and even generic HACCP being favoured in some sectors. There has been much focus in parts of the world on the use of prerequisite systems which 'design out' hazards and allow the HACCP Plan to control those hazards that are specific to the process. All of these issues (and more!) will be discussed.

We have set out in this second edition, using the same straightforward formula, to update and extend our practical advice on the use of HACCP systems.

An example HACCP study is again used to illustrate the theory in Chapter 6, but this time the modular approach, which is now more commonly used, is illustrated. A completely new set of HACCP case studies has been added to Appendix A, giving up-to-date ideas from HACCP practitioners.

Some of the changes have been stimulated by feedback from

readers. We welcome this input and would encourage readers of this edition to advise on any improvements for the future.

For those of you who have read and enjoyed the first edition, we hope that the second edition will not only bring you up to date with developments since 1994, but also provide food for thought and stimulate your ideas on HACCP as a major part of global food control. For those who are new to HACCP and this book, we trust that it will help you find your way to a successful HACCP system.

Sara Mortimore and *Carol Wallace*

Foreword

Since the 1994 publication of *HACCP: A practical approach*, many changes have occurred in the world of food safety. A number of driving forces have converged, focusing more attention on the proper management of food safety. These forces have prompted a revision and expansion of *HACCP: A practical approach*. Fortunately, the authors have been able to come forth with this timely revision of their most useful and excellent work.

Unquestionably, the most significant driving force for increased attention to food safety has been the continued surge in new foodborne pathogens and the related illness outbreaks. Micro-organisms such as *Salmonella typhimurium* DT104, antibiotic-resistant *Campylobacter jejuni*, *Cryptosporidium parvum* and *Cyclospora cayetanensis* were practically unknown in foods before 1994. However, most important in this regard has been the surge in major outbreaks of illness caused by *Escherichia coli* O157:H7 around the world. While it was originally found to be associated with dairy cattle, the ecological range of this pathogen is expanding. It is now a more frequent contaminant of raw animal foods and raw produce.

The surge in new foodborne pathogens and illnesses has led to unprecedented media attention to the safety of the global food supply. As a result, consumers are more aware of the potential problems and are demanding safer foods. Government regulatory agencies in many countries have responded by developing regulations for food safety. Many of these regulations require that the HACCP system of food safety be used in the production of food.

The remaining driving forces for food safety are not as obvious. As HACCP systems are implemented in food plants worldwide, it is important that these be done uniformly to facilitate the global sourcing of raw materials and the global distribution of foods. The need for a common understanding of HACCP principles and procedures is clear. Global companies are now requiring a uniform application of HACCP. New process technologies for the destruction of micro-organisms – high pressure, irradiation, pulsed electric fields, etc. – will need to be validated for their effectiveness in controlling micro-bial hazards. The revolutions in information technology and biotechnology are also beginning to impact food safety management. Some companies are beginning to monitor Critical Control Points electronically. The development of polymerase chain reaction technology may provide real-time monitoring for specific micro-organisms in a process.

The need for prerequisite programmes to support HACCP has also become apparent since 1994. Many of the procedures used for the management of HACCP, such as documentation, maintenance, and training, can also be used for the management of prerequisite programmes.

The purpose of this book is to provide the information necessary to establish and maintain a HACCP system. The authors have succeeded admirably in describing the application of the HACCP principles by using many examples, including a number of well-developed HACCP case studies.

HACCP: A practical approach has proved to be a valuable resource and has been very widely used in the four years since its publication. The revisions and expansion in this second edition, and its availability in three languages, will make it even more useful as a resource to help satisfy the global demands for improved food safety management. *HACCP: A practical approach*, second edition, will be equally helpful to food producers and regulators wordwide.

William H. Sperber
Senior Corporate Microbiologist
Corporate Food Safety
Cargill
Minneapolis, MN, USA

Acknowledgements

We are indebted to the following people for input into this book:

Dr W. H. Sperber   (Cargill Inc.)
Mr D. J. Phillips   (Pillsbury Europe)
Dr R. Evans   (Reading Scientific Services Ltd.)
Mrs S. Leaper   (CCFRA)
Mr A. P. Williams   (Williams & Neaves)

HACCP Case Study Authors:

Dr D. Worsfold   (Cardiff Business School)
Ms P. Nugent   (Cadbury Chocolate Canada Inc.)
Ms A. Sloan   (Foyle Meats)
Mr K. Y. Chong   (Singapore Productivity and
Mr A. Mak   Standards Board)
Mr P. de Jongh G.   (Cargill Venezuela)

Special thanks are extended to Ray Gibson and Colin Gutteridge at RSSL and Bill Newman at The Pillsbury Company for support throughout the project, and also to Di Amor (information support) and Mary Morris (typing).

Finally, we remain grateful to the following for their contributions to the first edition of *HACCP: a practical approach*, many of which remain in this second edition.

Mr A. L. Kyriakides   (J. Sainsbury plc)
Mr P. Catchpole   (Waterford Foods)
Mr P. Socket   (CDSC)
Mr A. W. Roberts, cartoonist
Mr L. Cosslett   (Britvic Soft Drinks)
Dr M. C. Easter   (Celsis International)
Mr D. B. Rudge   (Kerry Foods)

Disclaimer

The material in this book is presented after the exercise of care in its compilation, preparation and issue. However, it is provided without any liability whatsoever in its application and use.

The contents reflect the personal views of the authors and are not intended to represent those of The Pillsbury Company, Reading Scientific Services Limited, or their affiliates.

About this book

The purpose of this book is to explain what HACCP really is and what it can do for any food business. It will lead you through the accepted international approach to HACCP and will show you how to do it, from start to finish of the initial study, through to continuous maintenance of your system. The information given within the book may also be used as a basis for developing a HACCP training programme*.

You may have a number of reasons for wanting to know more about HACCP. If you are running a food business, you may already have decided that HACCP is the best control option available or you may have been asked to implement it by customers or regulatory authorities. You may be a regulatory authority and wish to enhance your understanding of the techniques. Alternatively, you may be studying the system as part of a broader course on food control or have noted the increasing recommendation for HACCP in legislation and guidelines.

Whatever your motivation for reading this book, there are a number of questions you will probably want to ask:

- What is HACCP?
- Where did it come from?
- Why should I use HACCP?
- Is HACCP available to everyone?
- How does HACCP help?
- What are the benefits?
- Are there any drawbacks?

* The contents of the book cover the requirements of the Royal Institute of Public Health and Hygiene HACCP Training Standard (RIPHH, 1995b). The book can therefore be used as the basis of a training programme designed to meet this standard.

- What are the principles of HACCP?
- Is it difficult?
- How do I do it?

You will find some brief answers in Chapter 1, with full explanations and examples showing 'how to do it' throughout the rest of the book.

The book has been designed to lead you through to a successful HACCP programme. It is based upon the international HACCP approach advocated by the Codex Alimentarius Committee on Food Hygiene (1993, 1997) and the National Advisory Committee on Microbiological Criteria for Foods (NACMCF, 1992, 1997) in the United States of America. The former is a committee of the WHO/FAO Codex Alimentarius commission.

It is recommended that you read the book from start to finish before going back over the 'doing' sections as you begin the HACCP process. As each of the HACCP techniques is explained, completed examples have been provided and there are additional case studies in Appendix A which can be used to look at different styles of application and documentation. Useful reference information on hazards and their control can be found in Chapter 4 and Appendix B.

1

An introduction to HACCP

HACCP is an abbreviation for the Hazard Analysis and Critical Control Point system. It is frequently written about and talked about at conferences and within companies, but is also often misunderstood and poorly applied in real situations. The HACCP concept has been around in the food industry for some time, yet it continues to be debated rigorously at international level. Developments in HACCP over the past 10 years or so have been fairly major, and some governments now see its implementation as a remedy for all of their country's food safety issues. In reality, use of the HACCP approach may well offer a practical and major contribution to the way forward, but only if the people charged with its implementation have the proper knowledge and expertise to apply it effectively. HACCP is a technique and needs people to operate it.

Those not familiar with HACCP often hold the misconceived belief that it is a difficult, complicated system which must be left to the experts, and can only be done in large companies with plentiful resources. True, you do need a certain level of expertise to carry out HACCP, but this expertise includes a thorough understanding of your products, raw materials and processes, along with an understanding of the factors that could cause a health risk to the consumer. The HACCP technique itself is a straightforward and logical system of control, based on the prevention of problems – a common-sense approach to food safety management. With good training, everyone ought to be able to understand the concept, given that it is based on sound reasoning. HACCP will be a key element of a broader product management system.

In brief, HACCP is applied by taking a number of straightforward steps:

- look at your process/product from start to finish;
- identify potential hazards and decide where they could occur in the process;
- put in controls and monitor them;
- write it all down and keep records;
- ensure that it continues to work effectively.

In this chapter, we will consider some of the most common questions asked by those who are new to HACCP.

1.1 Where did HACCP come from?

HACCP was developed originally as a microbiological safety system in the early days of the US manned space programme, as it was vital to ensure the safety of food for the astronauts. At that time, most food safety and quality systems were based on end-product testing, but it was realized that this could only fully assure safe products through testing 100% of the product, a method which obviously could not have worked as all product would have been used up! Instead it became clear that a preventative system was required which would give a high level of food safety assurance, and the HACCP system was born.

The original system was pioneered by The Pillsbury Company working alongside NASA and the US Army Laboratories at Natick. It was based on the engineering system, Failure, Mode and Effect Analysis (FMEA), which looks at what could potentially go wrong at each stage in an operation together with possible causes and the likely effect. Effective control mechanisms are then put in place to ensure that the potential failures are prevented from occurring.

Like FMEA, HACCP looks for hazards, or what could go wrong, but in the product-safety sense. Controls are then implemented to ensure that the product is safe and cannot cause harm to the consumer.

1.2 Why should I use HACCP?

HACCP is a proven system which, if properly applied, will give confidence that food safety is being managed effectively. It will enable you to focus on product safety as the top priority, and allow for planning to prevent things going wrong, rather than waiting for problems to occur before deciding how to control them.

Because HACCP is a recognized, effective method, it will give your customers confidence in the safety of your operation and will indicate that you are a professional company that takes its responsibilities seriously. HACCP will support you in demonstrating this under food safety and food hygiene legislation, and in many countries it is actually a legislative requirement.

Figure 1.1 'Origins of HACCP'.

When implementing a HACCP System, personnel from different disciplines across the company need to be involved, and this ensures that everyone has the same fundamental objective – to produce safe food. This is often otherwise difficult to achieve in the real world where there is constant pressure from a number of different areas, e.g. customer/commercial pressures, brand development, profitability, new product development, health and safety, environmental/green issues, headcount restriction, etc.

1.3 How does HACCP help?

The primary area where HACCP will help is in the processing of safe food. It helps people to make informed judgements on safety matters and removes bias, ensuring that the right personnel with the right training and experience are making decisions. As HACCP is a universal system, it can be passed on to your suppliers to assist in assuring raw material safety, and will also help to demonstrate effective food safety management through documented evidence which can be used in the event of litigation.

1.4 What are the benefits?

HACCP is the most effective method of maximizing product safety. It is a cost-effective system that targets resource to critical areas of processing, and in doing so reduces the risk of manufacturing and selling unsafe products.

Users of HACCP will almost certainly find that there are additional benefits in the area of product quality. This is primarily due to the increased awareness of hazards in general and the participation of people from all areas of the operation. Many of the mechanisms that are controlling safety are also controlling product quality.

1.5 Are there any drawbacks?

If HACCP is not properly applied, then it may not result in an effective control system. This may be due to improperly trained or untrained personnel not following the principles correctly; it may be that the outcome of the HACCP study is not implemented within the workplace; or it may be that the implemented system fails through lack of maintenance, e.g. if a company implements a system and stops there, paying little or no heed to changes that occur in the operation, then new hazards may be missed. The effectiveness may also be lost if the company carries out the hazard analysis and then tries to make its findings fit with existing controls. As we will see, HACCP is compatible with existing quality management systems but you must ensure that product safety is always given priority and that HACCP findings are not changed because they differ from existing operational limits.

Other problems may arise if HACCP is carried out by only one person, rather than a multi-disciplinary team, or where it is done at the corporate level with little or no input from the processing facility.

Some critics may say that HACCP is too narrow in that it focuses only on food safety; others say that it should only be used for microbiological safety. HACCP was designed for food safety and, as we have outlined, safety should always come first, but the HACCP techniques are flexible and can be applied to other areas such as product quality, work practices and to products outside the food industry. We will see much later how HACCP can be applied beyond food safety but the key issue is to make sure you don't try to do too much at once and end up either with a complicated system which is difficult to control, or with a system that covers many areas but lacks depth in each individual issue. As we will see in Chapter 9, it is better to apply the principles to each area separately, and then link together the management systems.

1.6 What are the Principles of HACCP?

The HACCP System consists of seven Principles which outline how to establish, implement and maintain a HACCP Plan for the operation under study. The HACCP Principles have international acceptance and details of this approach have been published by the Codex Alimentarius Commission (1993, 1997) and the National Advisory Committee on Microbiological Criteria for Foods (NACMCF, 1993, 1997).

We are now going to introduce a number of terms which may be unfamiliar to you. There is a glossary in Appendix C and an abbreviations list in Appendix D, and we will be discussing these again in full in Chapter 6 when we look at applying the Principles.

Principle 1

Conduct a hazard analysis. Prepare a list of steps in the process, identify where significant hazards could occur and describe the control measures.

Principle 1 describes where the HACCP Team should start. A Process Flow Diagram is put together detailing all the steps in the process, from incoming raw materials to finished product. When complete, the HACCP Team identifies all the hazards that could occur at each stage, establishes the risk to determine the significant hazards, and describes measures for their control. These may be existing or new control measures.

Principle 2

Determine the Critical Control Points (CCPs). When all the hazards and control measures have been described, the HACCP Team establishes the points where control is **critical** to assuring the safety of the product. These are the Critical Control Points or CCPs.

Principle 3

Establish Critical Limits for control measures associated with each identified CCP. The Critical Limits describe the difference between safe and unsafe product at the CCPs. They must involve a measurable parameter and may also be known as the absolute tolerance or safety limit for the CCP.

Principle 4

Establish a system to monitor control of the CCP. The HACCP Team should specify monitoring requirements for management of the CCP within its Critical Limits. This will involve specifying monitoring actions along with monitoring frequency and responsibility. In addition, procedures will need to be established to adjust the process and maintain control according to the monitoring results.

Principle 5

Establish the corrective actions to be taken when monitoring indicates that a particular CCP is not under control. Corrective action procedures and responsibilities for their implementation need to be specified. This will include action to bring the process back under control and action to deal with product manufactured while the process was out of control.

Principle 6

Establish procedures for verification to confirm that the HACCP System is working correctly. Verification procedures must be developed to maintain the HACCP System and ensure that it continues to work effectively.

Principle 7

Establish documentation concerning all procedures and records appropriate to these principles and their application. Records must be kept to demonstrate that the HACCP System is operating under control and that appropriate corrective action has been taken for any deviations from the Critical Limits. This will provide evidence of safe product manufacture.

1.7 Is HACCP applicable to everyone?

Yes, you may be a multinational food corporation who will be starting from within a sophisticated quality management system with documented procedures and well-defined practices, or a small manufacturer of goat's cheese in a shed on the farm, or a street vendor of ready-to-eat pizza slices. No matter, HACCP can be applied effectively to businesses at both ends of the spectrum.

HACCP is logical in its systematic assessment of all aspects of food safety from raw material sourcing through processing and distribution to final use by the consumer. Increasing numbers of terms are being used to describe the scope of the HACCP System. 'Farm to fork', 'stable to table' and 'gate to plate' are but a few and illustrate the fact that food safety control must encompass the entire food chain.

If we consider a simple supply chain model (Figure 1.2), we can see that there are various sectors within the food industry.

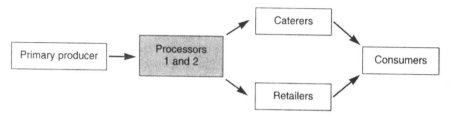

Figure 1.2 Supply chain model. Processors 1 includes primary converters, e.g. slaughter houses, chicken processors, sugar refiners, dairies. Processors 2 includes the secondary converters, e.g. finished-product manufacturers and packers.

This book will largely deal with HACCP application within the processors sector, but it is essential that HACCP is applied to the whole of the supply chain if food safety is to be assured. We will consider briefly how the Principles may be applied within the other areas:

1.7.1 Primary producers

These are the farmers, either raising livestock for the meat industry or the growers of the crops and vegetables that will be used by the processors in their conversion into finished products. The individual steps within the process can be assessed systematically for the potential for hazards to occur here, just as with any other area of the food-processing industry. Control measures can then be identified, and the control points that are critical to food safety identified. Critical Limits may be harder to identify, but here the farmer is often helped by legislative limits, for example in the case of herbicide and pesticide application.

Monitoring the CCPs can sometimes require some ingenuity. Staying with our example of herbicide and pesticide application, this may be done through the signing off of application record sheets or, when using aerial application, through use of regularly placed pieces of test paper across the land being sprayed, in order to record the spread of the application.

For primary producers there may be added difficulty in understanding the impact of their actions further down the supply chain. Yet for the processors it is almost impossible to anticipate what potential new hazards may arise at their stage in the supply chain, if they do not know what has occurred earlier on during primary production, or what may arise in the future. Also, an issue that may not appear to be a hazard on the farm may well have an impact further down the chain and require control measures to be implemented at the stage of the earlier primary process. For example, presenting animals for slaughter in an unfit state may increase the likelihood of *E. coli* contamination of the meat. Application of HACCP at the primary-producer stage is probably best done by use of a team approach. This could involve both the primary producers themselves, but also their customers (i.e. the processors, retailers and caterers). Some of the primary-producer trade federations in the UK are working to produce guideline support documents.

For further specific information, the *Assured Crop Production* guidelines (Campden and Chorleywood, 1996) and trade-specific codes of practice (e.g. Assured Produce Scheme, Checkmate International) will be of value to the crop-growing sector.

1.7.2 Caterers

Caterers, large and small, usually have a vast number of raw materials and a high turnover of staff. The principles of HACCP remain very relevant to this environment, however, and the implementation may begin with the use of a simple process flow diagram, as outlined in Figure 1.3.

With reasonable understanding of the basic rules of food safety and hygiene (Table 1.1), it should be possible to consider what might go wrong at each of the steps in the process. Adopting good hygiene practice within the catering environment at all times will be an essential ongoing requirement in order to ensure that the HACCP approach will add real value and that food safety management is enhanced.

Although not all caterers will have the in-depth technical knowledge to conduct what some might refer to as a 'real HACCP study', an attempt to understand and adopt the HACCP Principles should make significant improvement to the level of food safety control possible. The output of the studies may look less technical, the Critical Limits may not have been established through in-depth testing, but with a certain degree of external support, a simple but effective HACCP Plan can be put in place.

There is a catering case study in Appendix A but for further information *Assured Safe Catering* (HMSO, 1993) and FSIS *Food Safety in the Kitchen* (USDA, 1996a) may prove useful.

1.7.3 Retailers

As seen with the catering example, retailers should also be able to adopt HACCP to ensure that they sell safe food which the primary producers and processors have endeavoured to ensure reaches them in good condition. Correct temperature control and prevention of cross-contamination will be essential control measures in

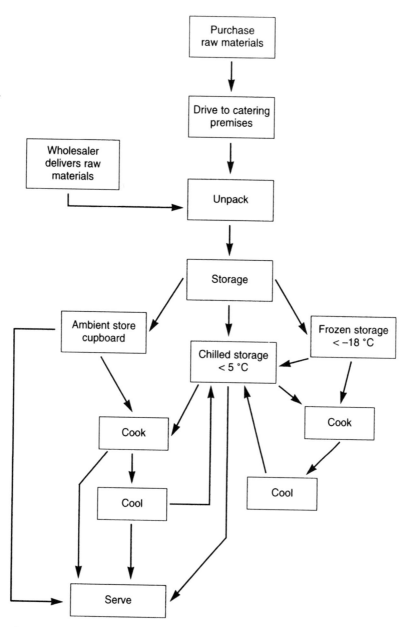

Figure 1.3 Simple catering process flow diagram.

both large and small premises. The HACCP application may be difficult for smaller shops, butchers for example, where both raw and cooked meat products have historically been sold by the same staff and from the same counter. In such examples changes to operating standards will almost certainly be required, but these can be

Table 1.1 Basic rules of food safety and hygiene within the kitchen

---

- Transport chilled and frozen food quickly between the shop and your premises. Place in fridge/freezer immediately you get back to the work place

- Keep the fridge below 5 °C and the freezer below −18 °C. Check this regularly with an accurate thermometer

- Check date codes and use all food within the recommended period

- Store cooked or ready-to-eat foods at the top of the fridge and raw foods at the bottom to avoid cross-contamination. Use separate fridges if possible

- Adopt good hygiene practice:

    – always wash hands before and after preparing food, after going to the toilet, handling rubbish, handling pets and handling raw foods

    – keep the environment clean, wash worktops and utensils between use for food which is raw and that which is cooked

- Cook food *thoroughly* and chill rapidly if not going to be used straight away

- Keep pets away from food, worktops and utensils. Keep doors and windows closed to prevent insect access

---

identified in a systematic way through use of the HACCP Principles. Like the caterers, for some of the smaller and independent retailers, the application is likely to be less technical, given the lower level of technical expertise available. However, the HACCP Principles, if truly understood and linked to good hygiene practices, should help to improve food safety control and hence significantly reduce risk. Effective training in both of these sectors is essential.

1.7.4 Consumers

This is a difficult area, as consumers do not necessarily have access to the education and training in food safety as does the food industry. HACCP techniques can be applied very successfully in the home environment (Griffiths and Worsfold, 1994), and to some extent there is much similarity between a domestic kitchen and that of the caterer. The simple process flow diagram shown in Figure 1.3 can easily be used for either situation, as can the control measures outlined in Table 1.1. It is important that consumers should take responsibility for storing, preparing and cooking all raw foods properly, rather than expecting all products to be completely free of micro-organisms at the point of purchase. However, it is equally vital that they are provided with correct usage instructions that allow adequate cooking to be carried out. Reliable sources of consumer education may exist, but, other than the product labels themselves, the process of obtaining this information is *ad hoc*, and sometimes the consumer is subjected to conflicting messages. Television cookery programmes are often very poor role models for good hygiene practice, and consumers are left to seek out literature from government bodies or retailers, if they want to know more (Mortimore, 1995).

HACCP can be used by everyone. The HACCP process itself is fairly logical and it is the hazard analysis step that can be the most difficult to get right without the proper expertise, i.e. knowledge of hazards and control measures. Determining Critical Limits can also cause problems, but the application of HACCP techniques outlined within the remainder of this book, while aimed primarily at the processor, can be interpreted for all sectors of the food industry.

2

Why use HACCP?

One of the first questions that will be asked within your organization is, 'Why use HACCP?' It is important to answer this question so that all personnel have:

- the same understanding of the motives behind the introduction of the system;
- commitment to developing an effective system.

In this chapter, we have endeavoured to cover some of the reasons for using the system – obviously for the management of food product safety, and in response to pressure from government and customers, but perhaps less obviously for business improvement.

2.1 Management of product safety

HACCP is a system of food control based on prevention. In identifying where the hazards are likely to occur in the process, we have the opportunity to put in place the measures needed to prevent those hazards occurring. This will facilitate the move towards a preventative quality assurance approach within a food business, reducing the traditional reliance on end-product inspection and testing.

All types of food safety hazards are considered as part of the HACCP System – biological, chemical and physical. Use of a HACCP System should therefore give the growers, manufacturers, caterers and retailers confidence that the food they provide is safe. Effective implementation of HACCP Systems can involve everyone in the company and each employee has a role to play. The culture that evolves through this approach makes it much simpler to progress to additional programmes such as quality improvement, productivity and cost reduction.

HACCP can, after the initial setting up of the system, be extremely cost effective. First, by building the controls into the process, failure can be identified at an early stage and therefore less finished product will be rejected at the end of the production line. Secondly, by identifying the Critical Control Points, a limited technical resource can be targeted at the management of these. Thirdly, the disciplines of applying HACCP are such that there is almost always going to be an improvement in product quality.

Consumer awareness of their right to purchase food that is safe has increased significantly over the past few years. Likewise, their awareness of quality failures, for example the presence of unwanted harmless physical contaminants, such as extraneous vegetable matter. Here the controls used to prevent the presence of a harmful contaminant, such as glass, are often likely also to prevent the occurrence of less harmful contaminants, therefore providing brand quality protection as well as consumer protection.

Foodborne disease continues to be one of the largest public health problems worldwide, and is also an important cause of reduced economic productivity. There are a number of reasons for this (Motarjemi *et al.*, 1996), including:

1. The proportion of the population who have increased susceptibility to foodborne illness is increasing, for example, the elderly, the immunocompromised and the malnourished.
2. Changing lifestyles have resulted in a number of changes to our eating habits:
   (a) more people now eat out, which has led to an increased demand for catering establishments of varying type;
   (b) many more women work outside the home and rely on processed foods for fast meal preparation; this has meant that knowledge of how to handle and prepare foods has decreased in recent years;
   (c) increased mass production of foods has increased the potential for larger numbers of consumers to be affected in the event of an outbreak of foodborne disease;
   (d) increased tourism has meant that people are exposed to foodborne hazards from other areas.

3. Emerging pathogens, such as verocytoxin-producing *E. coli*.
4. New technologies and processing methods.

The importance of the HACCP approach as a means of preventing foodborne illness has long been recognized by the World Health Organization and many governments worldwide.

HACCP was developed as a simple method of helping manufacturers assure the provision of safe food to the consumer, but many companies have only recently started to realize the full potential of the system. Let us consider some instances when HACCP could have prevented food safety incidents from occurring.

2.2 Examples of food safety incidents

As we have seen, the main reason for using HACCP is to manage food safety and prevent food safety incidents. Let's consider the possible consequences when food safety is not adequately managed.

When something goes wrong with a food product there may be localized or widespread illness and suffering, and the cost to the company concerned can be huge. Even when no illness has been caused, the discovery of safety hazards in a product intended for consumption can lead to prosecution for the company. Routine prosecutions often result from foreign material being discovered in food, but microbiological hazards have the potential to cause a much greater impact. Table 2.1 contains a number of food safety incidents that have occurred in recent years, along with an estimate of the likely cost. The true costs associated with such incidents are seldom documented, but where they have been established they can be shown to be substantial both to the industry and to society. For example, in the case of the *Salmonella napoli* outbreak in chocolate, the quoted costs relate solely to the health care costs and do not include the costs associated with withdrawing 2.5 million chocolate bars from the market.

A review by Socket (1991) of the Communicable Disease Surveillance Centre at Colindale in the UK reported that the costs associated with 10 salmonella incidents in catering establishments in the USA and Canada ranged from US$57 000 to US$700 000, and the direct costs only of five salmonella incidents in manufactured foods ranged from US$36 000 to US$62 million. It is also significant that the incidents listed in Table 2.1 involved both large and small companies and crossed international boundaries. Many of the companies involved received enormous publicity but for the wrong reasons, and not all are still in business. No company can afford to be a statistic in someone else's table.

An effective HACCP System is one way of preventing incidents such as the above from happening. It is a system where all hazards to food safety are identified and effective control mechanisms are put in place. The essential continued monitoring of these control mechanisms, and maintenance of the system, ensure that any potentially unsafe situations that occur are highlighted, and this means that the company can take appropriate steps to prevent a food safety incident.

Table 2.1 Implications of major food incidents

Year	Country	Food	Contaminant	Cause	Effect	Cost	References
1982	UK/Italy	Chocolate bars	*Salmonella napoli*	Cross-contamination	245 ill	£500k	Shapton (1989)
1985	USA	Mexican-style cheese	*Listeria monocytogenes*	Addition of raw milk	142 ill 47 deaths	Imprisonment $ millions	Shapton (1989)
1985	UK	Dried baby milk	*Salmonella ealing*	Cross-contamination	76 ill 1 death	£ millions	Shapton (1989)
1996/1997	Scotland	Cooked meats implicated (as yet unproven)	*E. coli* O157:H7	Cross-contamination or inadequate heat process suspected	20 deaths 501 affected	Unknown	The Pennington Group (1997)
1998	UK	Shell eggs	*Salmonella enteritidis*	Statement by government minister on the presence of salmonella in some eggs – but would be destroyed by adequate cooking	Public perceived eggs as unsafe	Immediate 60% reduction in sales. Overall cost >£3 million	North and Gorman (1998)

Year	Country	Product	Agent	Cause	Cases	Outcome	Reference
1994	USA	Ice-cream	*Salmonella enteritidis*	Cross-contamination from raw liquid egg transported in the same tank as pasteurized ice-cream mix	227 000 ill	Unreported	Hennessy et al. (1996)
1994	UK	Egg sandwiches	*Salmonella enteritidis*	Under-cooking	2 ill	Damages awarded £183 500 plus costs	Rogers (1994)
1989	UK	Hazelnut yoghurt	Botulinum toxin	Thermal process insufficient for reduced sugar hazelnut purée recipe	27 ill 1 death	£ millions. Whole yoghurt market depressed	Shapton (1989)
1993	World-wide	Various	Peanut allergen	Unknown consumption of product containing peanut	6 deaths	Unknown	Hourihane et al. (1995)

Table 2.1 *Continued*

Year	Country	Food	Contaminant	Cause	Effect	Cost	References
1996	Spain	Olive oil	Aniline dye	Denatured industrial rape-seed oil sold as edible olive oil	600 deaths 25 000 disabled	Unknown	WHO (1992)
1993	USA	Diet Pepsi-Cola	Syringes	Malicious contamination	Public confidence reduced; major crisis management programme involved	$40 million lost sales in one week	Pepsi-Cola Public Affairs 1993
1991	UK	Glass jar baby food	Glass/razor blades	Malicious contamination	Public confidence reduced. Total wet baby food market depressed	1988 £58m 1989 £45m/ US $4m	*The Grocer* (1991)
1990	France	Mineral water	Benzene	Filter not checked in 18 months	Worldwide recall, 160 million bottles destroyed	$79 million	Reuter (1990)

Year	Country	Food	Organism	Cause	Cases	Deaths	Reference
1997	USA	Ground beef	E.coli O157:H7	Raw ground beef is recognized as a potential source of this organism	25 million lb of meat recalled after health department investigation into outbreak	Unknown	Janofsky (1997)
1993	Germany	Potato crisps	Salmonella (90 serotypes isolated)	Spice mix contaminated	1000 cases (mainly children)	Unknown	Lehmacher et al. (1995)

2.3 Limitations of inspection and testing

So, what is wrong with what we are doing already – inspecting and testing? There are several points that you might like to consider here: first, 100% inspection where every single product manufactured is inspected. This would seem to be the ultimate approach to product safety, or would it? We often rely on inspection, particularly for finished products going down the production line, or ingredients during the weighing-up stage. Fruit and vegetables are good examples, where we look for physical contamination such as stalks, stones, leaves, insects, etc. Reasons why the technique is not as effective as we would like include the following: employees get distracted in the workplace by other activities going on around them, such as the noise of the production line or field environment, fellow workers talking about their holiday plans or what was on television the night before. The human attention span when carrying out tedious activities is short and 'hazards' could be easily missed during visual inspection. Because of this, people are often moved from task to task, in order to give some variety. However, this in itself brings problems along with line changes or shift changes; different personnel may be more aware of one hazard than another. Increasingly, electronic sensing techniques are being used to replace human input. These systems are more reliable but are not yet widely used except in large, more advanced food plants.

Of course, the main difficulty with a 100% inspection when it is applied to biological and chemical hazards is that there would be no product left to sell because biological and chemical testing is nearly always destructive. This leads us on to the use of sampling plans.

Many businesses randomly take a sample(s) from the production line. This can be daily, by batch, or even annually in the case of a seasonal vegetable or fruit crop. Statistically the chances of finding a hazard will be variable. Sampling products to detect a hazard relies on two key factors:

1. The ability to detect the hazard reliably with an appropriate analytical technique.
2. The ability to trap the hazard in the sample chosen for analysis.

Analytical methods for the detection of hazards vary in their sensitivity, specificity, reliability and reproducibility. The ability to trap a hazard in a sample is, in itself, dependent on a number of factors, including:

1. The distribution of the hazard in the batch.
2. The frequency at which the hazard occurs in the batch.

Figure 2.1 The limitations of inspection and testing.

Hazards distributed homogeneously within a batch at a high frequency are naturally more readily detectable than heterogeneously distributed hazards occurring at low frequencies.

For example, as illustrated in Table 2.2 , in a batch of milk powder contaminated with salmonella distributed evenly at a level of 5 cells/kg, a sampling plan involving testing 10 randomly selected samples, each of 25 g, would have a probability of detection of 71%. For powder contaminated at 1 cell/kg, the probability of detection using the same sampling plan would be only 22%.

This naturally assumes that the detection method is capable of recovering the salmonella serotype contaminating the batch. Few of the methods for salmonella detection would claim an ability to detect in excess of 90% of the 2000 serotypes, and most of the methods probably have a success rate of less than 75%. Therefore the low probability of 22% will be further reduced.

The probability of detecting a hazard distributed homogeneously in a batch is improved quite simply by increasing the overall quantity of the sample taken and is relatively unaffected by the number of samples taken. Therefore, 10 samples of 25 g would have the same probability of detection as one sample of 250 g.

However, in the majority of cases, hazards, particularly microbiological hazards, are distributed heterogeneously, often present in small clusters in a relatively small proportion of a batch. The probability of detecting a hazard distributed in this way is extremely low

Table 2.2 Detection probabilities – end-product testing, milk powder contaminated with salmonella

	Contamination rate	Number of random samples	Probability of detection*
Homogeneously contaminated	5 cells/kg	10	71%
	1 cell/kg	10	22%
Heterogeneously contaminated	5 cells/kg in 1% of batch	10	<2%
	10 000 cells/kg in 1% of batch	10	<15%

* Assuming detection test is 100% effective (most are <90%).

if low numbers of samples are taken. Using the example above (salmonella at 5 cells/kg), and assuming that the contamination is restricted to 1% of the batch, the probability of detecting the hazard by taking 10 samples of 25 g would be lower than 2%. Interestingly, even if the hazard occurred at high levels within 1% of the batch (10 000 salmonella cells per kg), the probability of detection would still be lower than 15%.

Such a situation cannot be rectified without recourse to a higher number of samples. In fact the probability of detecting the hazard in this scenario is greatly improved by merely taking more frequent samples from a batch, using a continuous sampling device. For example, if 100 g of the milk powder was removed from every tonne by a continuous sampler and a well-mixed sub-sample was tested (5 g from each tonne), the probability of detecting salmonella heterogeneously distributed at 5 cells/kg would increase from 2% to greater than 90%. However, even with exhaustive statistical sampling techniques, detection can never be absolute unless the entire batch is analysed, and in most cases few manufacturers understand or can afford to operate rigorous statistical sampling procedures.

In summary, if you look for hazards just by taking random samples, there is a high probability that they will not be detected.

2.4 External pressures

Increasingly, as HACCP becomes a regulatory requirement, this will be the main driver for its implementation. However, there will be other pressures, such as a real desire to improve consumer protection. Additionally, in recent years there has been an increasing amount of media interest in food safety issues, primarily focusing on the food-processing industry. It has been questioned whether the controls that are currently employed by the industry are adequate or in themselves may present increased risk to the consumer. We will go on to look at each of the main external driving forces for HACCP implementation.

2.4.1 Government

Government recognition of HACCP as the most effective means of managing food safety is increasing on a worldwide basis. The difficulty in focusing on specific pieces of legislation is that legislation is ever changing. However, in Europe one of the most powerful legal driving forces is the European Community Directive 93/43 EC (1993) on the hygiene of foodstuffs. The Directive, while not using the precise wording of Codex Alimentarius or NACMCF, in Article 3 states that 'food business operators shall identify any step in their activities critical to ensuring food safety and ensure that adequate safety procedures are identified, implemented, maintained, and reviewed'. In essence the Directive lists the first six principles required to develop the system of HACCP, and can be interpreted in virtually the same way as Codex/NACMCF, with the exception of any specific reference to record keeping. The Directive dictates that competent authorities shall carry out controls to ensure that this is being complied with by food businesses; obviously evidence of compliance will be required, i.e. records. Where failure to comply results in risks to the safety or wholesomeness of foodstuffs, appropriate measures shall be taken which may extend to the withdrawal and/or destruction of the foodstuff or to the closure of the business for an appropriate period of time.

The adoption of the Directive means that all food businesses throughout Europe are strongly recommended to use the HACCP approach, in that it will enable them to meet the requirements of the legislation. It also, more specifically, means that food businesses that are certified to the international quality standard ISO 9000 will be forced to include HACCP within the scope of their Quality Management Systems, as under this standard all relevant legislation must be complied with in full.

In the UK, the statutory defence of Due Diligence contained within the Food Safety Act (1990) requires that the person proves that he took 'all reasonable precautions and exercised all due diligence to avoid the commission of the offence by himself or by a person under his control'. A defendant using this defence in case of litigation would certainly have a stronger case if it could be proved that HACCP was in place and working.

The UK Government has recognized HACCP in several specific

reports and no doubt this recognition will increase. In New Zealand, the Ministry of Agriculture is in the process of making HACCP mandatory for all food producers. In the USA the HACCP techniques were used to identify the controls specified in the Low Acid Canned Food Regulations. The US Department of Agriculture has decreed that HACCP programmes will be required for all meat- and poultry-processing facilities, beginning in 1998. It is also required by law in the area of seafood inspection and processing (Federal Register, 1995, 1996). The US Food and Drug Administration is considering additional HACCP regulations on an industry-by-industry basis. At the time of writing a rule for fruit-juice processing is nearing completion. The trend seems to indicate that HACCP will eventually be mandatory not only for all US food processing facilities but also all food processors who are exporting into the United States from anywhere else in the world. In Canada, manufacturers of high-risk food products were required to have HACCP plans in place by the end of 1991.

In summary, it is clear that international legislation is moving more and more towards making HACCP a mandatory requirement for the food industry. Key indicators include the legal requirement for use of HACCP in specific sectors of the food industry and the strong recommendation from many governments through direc-tives and food safety reports and surveys.

2.4.2 Customers and consumers

Within the food industry, the safety of our products must, without question, be considered top priority. That food is 'safe' is often an unwritten requirement of many customer specifications. It goes without saying and, unlike many of the other attributes of the prod-uct (appearance, taste, cost), it is not negotiable. Consumers expect safe food and we in the food industry have a responsibility to meet their expectations.

While the end consumer may not know what HACCP means, those of you who are supplying to retail and food-service customers are most likely to be asked to implement a HACCP System. Indeed, it may have been one of the main driving forces behind the purchase of this book. For both the retailer and caterer the customer is the end of the supply chain, i.e. the consumer of the food. For the grower and food manufacturer, quite likely the customer is a caterer, a retailer or another manufacturer. Whatever the situation, customers want to be confident that the food being purchased is safe. They want to have confidence in their supplier.

HACCP is an excellent way of assuring the safety of the food, because not only must it be carried out and verified, but also it must be maintained. Gone are the days when a customer inspection meant a walk around the factory to check hygiene and housekeeping, followed by a pleasant lunch. A crucial factor in any supplier inspection these days is an assessment of the competence of the management. Larger customers are likely to issue their own 'Codes of Practice' which almost certainly will include the requirement for a HACCP System to be in place. An effective HACCP System can go a long way in demonstrating to the customer that the supplier is managing the food safety hazards.

In some countries, particularly the UK, there is an increasing trend towards third-party audit of suppliers. This tends to be carried out either as a part replacement or as an enhancement of the customer's own inspection activities. There can be a benefit to the supplier in that the audit bodies often have considerable experience within the industry and can provide a useful challenge to the HACCP System. One of the main reasons for the change in supplier/customer relationships in the UK has been the Food Safety Act (1990) according to which a company must prove that it 'took all reasonable precautions and acted with all due diligence'. The supplier inspection is very much about this – establishing faith in the suppliers.

In turn you may want to consider asking your own suppliers, particularly of high-risk materials, to implement a HACCP System. Perhaps you could give a 'preferred supplier status' to those who have done so. Your suppliers may then be encouraged to pass the disciplines back through the supply chain to their suppliers, and so on.

No one wants to be buying-in a problem. If a food safety incident was attributed to your product, but was eventually traced to an ingredient, would it be you or your supplier who was held responsible? It may turn out to be the supplier's fault, but what damage will have been done to your business in the meantime if the media have taken an interest and your brand is involved?

Where does the consumer feature with respect to food safety control? As mentioned earlier, the scope of HACCP can cover the entire food supply chain, right down to the kitchen of the final consumer (Griffiths and Worsfold, 1994). The consumer has typically played the role of lobbyist in demanding assurance of safe food, and hence has been a driver for implementation of food safety management systems by the industry. However, consumer perception of risk severity does not necessarily always correlate with that of the food industry experts. Table 2.3 indicates that consumers classify pesticides, for example, as a much greater risk to their health than natural toxicants. These perceptions are important for a number of reasons. Referring back to HACCP application in the consumer sector (section 1.7.4), clearly, if consumers do not perceive

Table 2.3 Risk perspective of some chemicals in food

	Greatest risk
Scientist's perception:	**Consumer's perception:**
1. natural toxicants	1. pesticide residues; food additives
2. chemicals migrating from food-contact materials; pesticides; drug residues	2. drug residues
	3. chemicals migrating from food-contact materials
3. food additives	4. natural toxicants
	Least risk

Source: R. Evans, personal communication

themselves as being very highly exposed to a food safety risk, then they aren't going to adopt the necessary control measures.

Although not within the scope of this publication, the authors recognize that food hygiene education of the consumer is a vital element in prevention of foodborne illness. Education should include the principles of good hygiene practice in the home, how to prevent cross-contamination, and the importance of temperature in controlling microbiological food safety. Some governments are starting to work with industry and trade organizations in acknowledgement that improved understanding and consumer ownership of preventative control measures will result in a decrease in the number of food poisoning outbreaks.

Additionally, the food industry is a major employer and the possibility of potential employees having a greater awareness of basic good hygiene practice would be a real advantage.

2.4.3 Enforcement authorities

The role of the enforcement authority is to ensure that legislation is being complied with correctly. In the UK this is the responsibility of the Local Authority Environmental Health Departments, but there are equivalent or similar bodies elsewhere, e.g. the Food Safety Inspectorate in the USA.

Environmental Health Officers (EHOs) in the UK are provided with Codes of Practice by the Department of Health. These offer guidance on the interpretation and enforcement of specifically *The Food Safety Act (1990) Code of Practice No. 9: Food Hygiene Inspections* (HMSO, 1997) makes reference to the use of Hazard Analysis Systems in relation to inspection of food businesses. Either the EHO would review a HACCP System already in place, i.e. evaluate the understanding, or instead use the Hazard Analysis techniques to determine which hazards will need to be controlled in order to ensure safe food production. The latter approach may be used in smaller businesses where fully documented HACCP Systems are not being implemented by the company itself.

In Australia and New Zealand, an internationally acceptable standard is being developed by the Joint Australia and New Zealand Standards Board. This will be used eventually by third-party auditors for HACCP certification. Similar developments are under way elsewhere in the world (Majewski, 1997).

2.4.4 Media

Most companies are aware of the power of the media but perhaps feel complacent when it comes to their own businesses, thinking 'it will never happen to us', 'we are in control', etc. Food safety scares have become big business; the media are always looking for a good story and consumers feel encouraged to go to the press, lured by both the publicity and the cash rewards.

Sometimes the issues may be very real, but not always. If a consumer goes to the press you will need to have evidence in order to answer the claims made against you. This is particularly important if the consumer has falsified claims and the police are drawn into the case. Fully documented evidence, through HACCP records which have been efficiently maintained, is essential. Someone within the company who is trained in media handling and an effective incident management system could be vital in ensuring that the company remains in business and the risk to the public is minimized in the event of an incident occurring.

2.4.5 International standardization

Improvements in distribution technology have contributed to the increased globalization of food trade. The intent of the Codex Alimentarius Commission (CAC) is to facilitate international trade by providing a documented standard that is based on improved consumer protection and fair trade practices (Hathaway, 1995). The CAC is able to influence food regulation worldwide and utilizes the food safety best practice standards adopted by member governments in drawing up the Codex Alimentarius standards.

Since the early days in Pillsbury, HACCP Principles have become accepted internationally, and the common understanding has been assisted by the publication of the seven HACCP Principles (outlined in Chapter 1) within the Codex (1993 and 1997 update) documents. From these documents, many manufacturing companies, committees, consultancy groups and food research associations, large and small, have taken a lead.

This has steered the way towards harmonization in HACCP worldwide and has been helpful with respect to international trade. As a result of the completion of the General Agreement on Tariffs and Trades (GATT) Uruguay Round and the establishment of the World Trade Organization (WTO) in January 1995, mutual agreement of the standards of each trading partner's country and/or the equivalence of food safety systems must occur before trade can proceed. While many countries are in the process of re-evaluating their food safety policies (Majewski, 1997), use of the Codex HACCP Principles as the international standard means that the HACCP System implemented by one company is based on the same Principles as those installed by its competitor, wherever in the world that competitor happens to be based. What remains then is the detailed interpretation of the Principles which, to date, the CAC has not taken on as a role. There continues to be considerable debate at international level regarding interpretation and implementation of HACCP.

This book has the objective of contributing to a common understanding of HACCP and is based on the Codex approach.

2.5 Prioritization for improvement

HACCP is a system of managing food safety, but once you have learnt the technique it can be applied to other aspects of the business. One of the main benefits in the early stages of implementation is its help in setting priorities. Mistakenly, many people feel that HACCP can only be used by mature businesses who have Good Manufacturing Practices and even Quality Management Systems such as ISO 9000 series already in place. It should be argued fiercely that HACCP is especially important for those businesses who do not fall into this category. HACCP can be used to prioritize areas for improvement.

By systematically analysing the hazards at each stage in any food production chain and determining at which points control is critical to food safety, you can see whether you already have these controls in place or not. The same discipline can be used to determine where control is crucial to end-product quality (appearance, taste), shelf life (what factors are important to control spoilage?) and legality (e.g. weight control). A HACCP study can also be used to assess where priorities lie in terms of Supplier Quality Assurance. For all materials an agreed specification will be needed, but how do you know which suppliers to visit in order to assess their control of potential hazards, whether a certificate of analysis will be necessary, and when to sample and test at your own factory? Again, by determining which of your raw materials are themselves Critical Control Points (i.e. control is critical to the safety of the end product), you can start to prioritize activity and make effective use of the resource available.

2.6 Meeting food safety objectives

Key food safety objectives may include:

- producing a safe product every time;
- providing evidence of safe production and handling of food products; this is particularly useful during regulatory inspection or prosecution;
- having confidence in your product and thereby ensuring that customers have confidence in your ability;
- satisfying a customer request for HACCP to an international standard;
- compliance with regulatory requirements.

Additional objectives may include:

- involving personnel from all disciplines and at all levels in HACCP implementation, the management of food safety becoming everyone's responsibility.

Use of the HACCP system is a structured and systematic way of ensuring that these food safety objectives are achieved.

2.7 The benefits

We discussed a number of the benefits of HACCP in Chapter 1. In order to describe these fully you need to be clear about the purpose for which *you* are implementing HACCP. Why are the results needed? What are the benefits to be gained? Are they only to be measured in terms of product quality or are there non-product benefits, such as improved process understanding or improved team working? Are any financial benefits likely to accrue from effective HACCP implementation? Whatever the benefits and purposes are, it is helpful to record them, as they can then be used throughout the development of the programme as a reference point to help you prioritize or decide direction.

3

Preparing for HACCP

When the decision is taken to use HACCP within a company, there is often the inclination to charge ahead and start doing something without taking the time to consider the best approach for the company.

Before you get too far down the road, you need to consider where you are now and where you need to get to in terms of food safety management. In this chapter we will look at how to plan the application and implementation of the HACCP Principles. This includes an indication of how to prepare the way and plan the project, how to evaluate and build effective support systems, and how to identify and train the types and numbers of people required to establish and manage an effective system.

The way to implement the HACCP Principles may at first seem obvious, particularly after a training course; but have you considered the various alternatives in terms of the structure of the HACCP System? A degree of forethought now will pay benefits later on, not least because it will give other people within the organization the chance to visualize what you are about to do and allow them to make relevant and valuable contributions to your effort. This is more likely to result in the implementation of a successful system, which gains commitment for ongoing development and further improvement throughout the business.

Any company that is new to the HACCP techniques will go through four key stages to obtain an effective system (Figure 3.1).

In this chapter we will be discussing key stage one – Planning and Preparation (Figure 3.2). This is where the foundations are laid and it is important to take time here to:

- ensure that the appropriate people are identified and trained;
- establish what support systems are already in place and what needs to be developed;

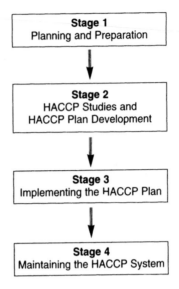

Figure 3.1 The key stages of HACCP.

- consider the most appropriate structure for the HACCP System;
- plan the entire project, including a realistic timetable for development and implementation of the HACCP Plan.

The first thing to do is to consider what you are trying to achieve. This can be done by reading the rest of the book to gain a good understanding of HACCP before starting out. The path you take to a fully implemented HACCP System will then depend on where you are starting from and the maturity of your existing systems.

3.1 Preparing the way – personnel and training

3.1.1 Personnel resources

HACCP is carried out by people. If the people are not properly experienced and trained then the resulting HACCP System is likely to be ineffective and unsound. In this section we will discuss the people who need to be involved and assess their training requirements. Also we will help you to identify the right experts for your company.

(a) Senior management

Early involvement of senior management is fundamental to the effective implementation of HACCP. Real commitment can only be achieved if there is complete understanding of what HACCP

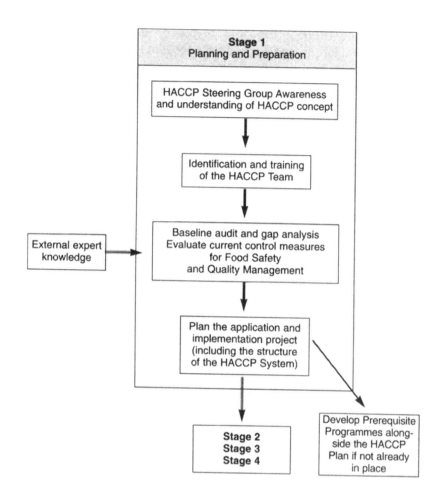

Figure 3.2 HACCP Key Stage one – Planning and Preparation

actually is, what benefits it can offer to the company, what is really involved and what resources will be required. This understanding will be achieved not only by reading books such as this one but also by attending a HACCP briefing session, as a senior management group. This may be undertaken by a reliable consultant if there is no one able to do it internally. Open discussion should then follow, with the end result that the decision to go ahead is given full support by the board.

Senior management from all disciplines must be encouraged to show their support actively and be unanimous in their approach. It would be a pity if credibility was lost through the Sales Director continuing to make rash promises to the customer: 'Yes, we can develop and produce this completely untried and untested high-risk product for you within 3 days, no problem', or through the Engineering Manager purchasing equipment that may be unable to achieve the process criteria needed to make a safe product or be cleaned properly.

Identification and training of a HACCP steering group will provide a valuable and visible support to the implementation of HACCP. Functional Senior Managers plus the HACCP Team Leader are vital to form this group.

(b) The HACCP Team

It is important that HACCP is not carried out by one person alone but is the result of a multi-disciplinary team effort – the HACCP Team. The second preparatory activity is to identify and train the HACCP team. It is recommended that as a minimum the core HACCP Team consists of experts ('expert' meaning having knowledge and experience) from the following areas:

1. **Quality Assurance/Technical** – providing expertise in microbiological, chemical and physical hazards, an understanding of the risks, and knowledge of measures that can be taken to control the hazards.
2. **Operations or Production** – the person who has responsibility for and detailed knowledge of the day-to-day operational activities required in order to produce the product.
3. **Engineering** – able to provide a working knowledge of process equipment and environment with respect to hygienic design and process capability.
4. **Additional expertise** – may be provided both from within the company and from external consultancies. The following areas should be considered:

   • Supplier Quality Assurance – essential in providing details of supplier activities and in assessment of hazard and risk associated with raw materials. The person responsible for auditing suppliers will have a broad knowledge of best practices gained through observing a wide range of manufacturing operations.
   • Research and Development – if the company is one where new products and process development is a continuous activity, then input from this area will be essential. Early involvement at the product/process concept stage could prove invaluable.
   • Distribution – for expert knowledge of storage and handling throughout the distribution chain. This is particularly important if distribution conditions, e.g. strict temperature control, are essential to product safety.

- Purchasing – could be useful in a company involved in purchasing factored goods for onward sale or in a food service operation, or where technical and purchasing roles are combined. The participation of purchasing personnel will mean that they are made fully aware of the risks associated with particular products or raw materials and can assist with early communication of a proposed change in supplier.
- Microbiologist – if the company has its own microbiologists, then their expert knowledge is likely to be needed on the HACCP Team. Many smaller companies do not have this option and, where microbiological hazards require consideration, they should identify a source of expert help from either a Food Research Association or from a reputable consultancy or analytical laboratory.
- Toxicologist – likely to be located in a Food Research Association, consulting analytical laboratory or university. A toxicologist may be needed particularly for knowledge of chemical hazards and methods for monitoring and control.
- Statistical Process Control (SPC) – there are many courses now available which will be sufficient to give members of the HACCP Team or their colleagues enough knowledge to carry out basic SPC studies on their process operations. This will be important in assessing whether a process is capable of consistently achieving the control parameters necessary to control safety. In some instances, however, it may be advisable to have an external expert join the HACCP Team as a temporarily co-opted member. This would be useful when setting up sampling plans or for a more detailed assessment of process control data.
- HACCP 'experts' – it may be appropriate initially to co-opt an external 'expert' in HACCP onto the HACCP Team. This may be useful in helping the company team to keep on the right track and become familiar with the HACCP System. It could also be extremely important in helping the company to determine whether they have got the right people on the team and as an early assessment of whether the initial HACCP studies are correct.

If the company does not already use team working, it may be difficult initially for individuals to adjust to this approach. It should be explained that as a team effort the HACCP Study will have input from a much greater diversity of knowledge, skills and experience, far beyond that of any one individual. The team is made up of people with a real working knowledge of what happens in each area and therefore any processes that cross over departments can be tackled more accurately. You should also consider that HACCP studies may well result in recommendations for changes to processes and products and even on occasions capital expenditure. These recommendations are far more likely to be accepted by senior management if they are supported by knowledgeable people across all disciplines within the company.

We have now considered the disciplines required within the team and, in summary, it should be emphasized that expert judgement is essential in assessment of hazards and risks. What else is important with respect to the type of people involved in the team? Personal attributes will include:

1. Being able to evaluate data in a logical manner using expertise within the team and perhaps using published data for comparison.
2. Being able to analyse problems effectively and solve them permanently, treating the cause not the symptom of the problem.
3. Being creative by looking outside the team and the company for ideas.
4. Being able to get things done and make recommendations happen.
5. Communication skills. The HACCP Team will need to be able to communicate effectively both internally within the team and externally, across all levels of the company.
6. Leadership abilities. Leadership skills of some degree will be useful in all members of the team. After all, they are leading the company in its HACCP approach to food safety management. It is recommended that one member of the team is appointed to HACCP Team Leader. This is often the QA Manager but consider carefully what the leadership of a team entails. Your Personnel or Human Resources department may be helpful in identifying suitable courses for development of these skills if they are not already sufficient.

The Team Leader will have a key role in the success of the HACCP System and he or she is likely to become the company HACCP expert and be regarded as such. In the leadership role the Team Leader will be responsible for ensuring that:

- the team members have sufficient breadth of knowledge and expertise;
- their personal attributes are taken into account;
- individual training and development needs are recognized;
- the team and work tasks are organized adequately;
- time is made available for reviewing progress on an ongoing basis;
- all skills, resources, knowledge and information needed for the HACCP System are available either from within the company or through identifying useful external contacts.

The behaviour within the team must be supportive, encouraging all members to participate. With all team members fully committed to producing and maintaining an effective HACCP System there should be no time for arguments or internal politics.

HACCP Team size

Within the HACCP Team itself, consider the range of disciplines required. In smaller companies the same person may be responsible for both QA and Operations. In terms of team size, four to six people is a good range. This is small enough for communication not to be a problem but large enough to be able to designate specific tasks.

In large organizations there may be more than one HACCP Team. It was stressed earlier that the members of the team must have a good working knowledge of what actually happens in practice. In large companies the 'experts' and senior people in the three main disciplines of QA, Production and Engineering may not be close enough to the operation. It may then be more effective to have a series of smaller departmental factory teams, still made up of the three main disciplines but at a less senior level. The departmental teams then carry out the HACCP Study for their own areas and, when satisfied with the resulting HACCP Plan, pass it up to a more senior level for approval. This ensures that the true working knowledge of activities is captured and subsequently reviewed by appropriate experts in each area. This approach is also common when HACCP is applied to the process in modular form (see section 3.3.1 (b)). An example of this could be represented diagrammatically, as in Figure 3.3.

(c) Additional personnel

In addition to the HACCP Team and senior management, personnel throughout the operation will need to be involved. This will include line supervisors, operators, incoming goods inspectors, cooks and point of sale personnel. It is likely that these people will be involved later on, when HACCP moves into the implementation phase. It is important that they, too, are fully briefed on their role within the system, particularly if they are monitors of the controls critical to food safety. We will discuss this in more depth in Chapter 7.

The numbers of people needed in addition to the HACCP Team will be dependent upon the type of operation and number of controls that need to be monitored. There should always be a sufficient number of people to ensure that the critical points are monitored effectively and that records are reviewed.

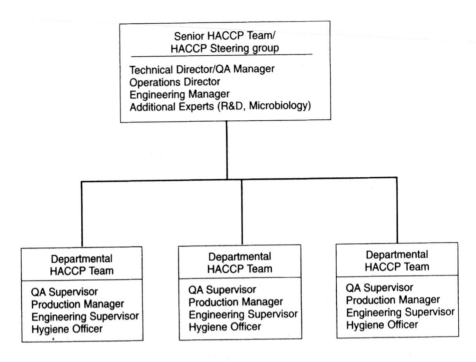

Figure 3.3 Example of HACCP Team structure in a large organization.

3.1.2 *What are the training requirements?*

HACCP is only going to be effective as a means of managing food safety if the people responsible for it are competent. As a result, training becomes the single most important element in setting up a HACCP System. Training not only provides the technical skills required in implementing HACCP, it also helps in changing attitudes of people. In this section we will explore the training requirements for HACCP Teams. Training for other personnel is covered in Chapter 7.

In our experience, a number of key competencies are required of HACCP Teams and a balance of these attributes throughout the team is necessary.

The training of these people is an investment and should be taken seriously. You should realize that the HACCP Team members may need to be provided with many additional support skills in addition to the specific HACCP knowledge (such as project planning, statistical process control, audit skills, team working, etc.).

Tables 3.1 and 3.2 outline baseline training requirements for HACCP. A combination of these may be required for each HACCP Team member.

Table 3.1 outlines a training programme which would usually be required for all HACCP Team members. This level of training offers a good introduction to the HACCP Principles and their application and may often be provided by an external organization. When choosing a course, it is vital to check that it covers all the required elements and gives practical, 'hands-on' experience as well as covering the theory (Figure 3.4). It is also worth checking whether the course has been evaluated successfully against a Training Standard, such as the WHO-endorsed Training Standard in HACCP Principles (RIPHH, 1995b) in the UK. If you have internal training experience, you may wish to use this book to develop your own training material, or to purchase the sister publication, *A Practical Approach to HACCP, Training Programme* (Mortimore and Wallace, 1997).

Figure 3.4 How do you draw a Process Flow Diagram?

Table 3.1 Baseline Training Programme for HACCP Team Members*

Course Outline

Module	Learning Objectives
1. Introduction	a. Course logistics b. Objectives
2. Relationship of good manufacturing/hygienic practice (GMP/GHP) to HACCP	a. Understanding that, prior to implementing HACCP, effective GMPs/GHPs as defined in the Codex General Principles of Food Hygiene must be in place b. Understanding consequences for HACCP implementation, if GMPs/GHPs are not in place
3. Overview of HACCP	a. Understanding the HACCP approach to food safety and that it differs from traditional approaches b. Learning about the advantages of the HACCP System c. Understanding that without management commitment, HACCP cannot be implemented effectively on a sustained basis d. Understanding that there are seven principles to HACCP and that they are easy to comprehend
4. Discussion of foodborne hazards	a. Understanding that there are hazards associated with foods which, if left uncontrolled, can injure consumers b. Learning what a hazard is and the three different categories of hazards, i.e. biological, chemical or physical c. Understanding that there are ways of addressing hazards (control measures) to prevent, reduce or minimize hazards associated with foods

5. Preliminary steps to develop a HACCP Plan

   a. Understanding the importance of HACCP Team formation and activities: i.e. the advantages of a multi-disciplinary team including – but not limited to – the following skills: personnel in production, quality assurance, engineering

   b. Understanding the need for a person, trained in HACCP, to lead the HACCP Team and to be skilled in meeting management

   c. Describing products to be covered by HACCP and their intended use

   d. Understanding the importance and use of process flow diagrams and their verification for accuracy

6. Conducting a hazard analysis

   a. Identifying all potential hazards and points where they enter the process/food or can be enhanced during the process

   b. Identifying significant hazards relative to severity and likelihood of occurrence

   c. Identify appropriate control measures for significant hazards

   d. Documenting the rationale for hazard selection for future use

7. Identifying CCPs

   a. Understanding a CP and a CCP and their difference

   b. Understanding relationship between an identified hazard and a CCP

   c. Utilizing a decision tree to identify CCPs and to address all significant hazards

8. Determining critical limits

   a. Understanding critical limits and operational limits and determining each one

   b. Selecting critical limits which have relevance to the safety of a product but which are not unnecessarily restrictive

   c. Documenting the rationale for the selection of critical limits for future use

9. Establishing a monitoring system

   a. Identifying how monitoring is to be conducted

   b. Identifying who will be responsible for monitoring activities

   c. Determining the necessary frequency for taking measurements

   d. Identifying where measurements will be taken

   e. Understanding the importance of monitoring for control purposes

Table 3.1 *continued*

Module	Learning Objectives
10. Establishing corrective actions	a. Learning to devise effective corrective actions b. Understanding the need for appropriate documentation of corrective actions
11. Establishing verification procedures	a. Learning the importance of verification to support HACCP and help assure its long-term viability in an establishment b. Learning different activities that can be conducted as part of verification
12. Establishing record keeping and documentation	a. Learning the type of records needed to document HACCP activities b. Learning the importance of record keeping for determining the effectiveness of the HACCP System c. Learning what information should be included in records d. Learning the importance of reviewing and signing records
13. Management of HACCP	a. Understanding the importance of management support b. Understanding the importance of operational units in supporting a HACCP System, e.g. supervisors, quality control/quality assurance, technical c. Understanding the importance of standard operating procedures and how they are used d. Understanding how to assemble and maintain a HACCP Plan e. Understanding the composition, role and importance of the HACCP Team
14. Implementation strategies	a. Training Programmes (GMPs – HACCP Plan support) b. Preparation of standard operating procedure c. Transfer of ownership (responsibility for food safety) d. Gaining management support
15. Maintenance of HACCP Plans	a. Understanding that plant management is responsible for ongoing safety of food, implementation and management of the HACCP Plan b. Learning to review and revalidate the HACCP Plan

16. HACCP Plan
   a. Learning the constituents of the HACCP Plan and supporting documents
   b. Standard operating procedures (sample record forms, sampling plans, monitoring procedures, calibration procedures)

17. Auditing
   a. Understanding the objective of auditing
   b. Learning about those who may perform auditing (food companies, third parties and/or government officials) and their role
   c. Learning what is audited: records, procedures and products
   d. Learning how to plan the frequency of auditing (i.e. considering the risk factor, compliance, etc.)
   e. Learning how to audit in order to ensure ongoing support for HACCP, document compliance with the HACCP Plan, and to determine if the Plan needs to be changed

* Adapted from WHO (1995).

Table 3.2 Suggested sources of additional HACCP Team skills and knowledge

Skill/knowledge	Means of providing it
1. Principles and techniques of HACCP	In addition to the training described in Table 3.1, reference books and scientific papers: ICMSF (1988); Pierson and Corlett (1992); Mortimere and Wallace (1994, 1998); Campden and Chorleywood Food Research Association (1997)
2. Understanding of the types of hazards that could occur and methods of control. For example, in relation to foodborne pathogens this should include the frequency and extent of their occurrence in different foods; the severity and likelihood of transmitting foodborne pathogens and toxins through different foods; the means of and influence of contamination of all types and elimination or reduction by processing and procedures, i.e. the control measures	With a good mix of disciplines on the team this area should be covered, provided the team members, among them, have both academic backgrounds in microbiology or food science-related subjects, and sufficient food industry experience. Useful courses in understanding hazards are provided by many training organizations if a refresher is needed. Additionally, an advanced qualification in food hygiene is highly recommended. In the UK this is provided by several reputable bodies such as the Royal Institute of Public Health and Hygiene, the Chartered Institute of Environmental Health and the Royal Society of Health Use of hazard databases available through food RAs and universities Use of the Internet Use of reference books: ICMSF (1980, 1986, 1996); Sprenger (1995)
3. Detailed knowledge of Good Manufacturing Practices	Essential food industry experience as above Reference books: IFST (1991); Shapton and Shapton (1991)
4. Team-working skills, including communication skills (especially important if this is a new way of working for most team members)	Personnel department may be able to assist with some in-house team-building training for the HACCP Team External team-building courses are available, often lasting about 5 days

5. Project planning and management skills (the HACCP implementation project may have a separate Project Manager but, if the HACCP team itself is responsible, this skill will be invaluable

External courses run by management consultancies or training organizations
Use of an on-site consultant in the early stages
Reference books: Bird (1992); Brown (1992); Oates (1993)

6. Auditor training – essential for the verification of the flow diagram and HACCP Plan

A Quality Management Systems auditor course is recommended (internal auditors level is sufficient, which usually lasts 2 days). These can be run on your own site if numbers justify. Available from ISO 9000 assessment bodies, professional institutes or training organizations
Reference books: Chesworth (1997)

7. Statistic Process Control (a working knowledge in order to make valid process capability assessment and data handling)

External management consultancy groups who often provide training packages
Reference books: Rowntree (1981); Price (1984)

8. Problem-solving techniques – in order to tackle recurring problems in a structured way and ensure that permanent solutions are found. Can be very useful in learning how to draw process flow diagrams and in handling data

Training packages can be purchased from management and training consultancy groups
Courses are also available through the above. Recommend an on-site session tailored to the need (HACCP) in order for it to be really understood and applied after the event

Table 3.2 *continued*

Skill/knowledge	Means of providing it
9. Trainer training skills – essential if HACCP training is to be carried out in-house	Food industry courses are now being run by many of the food training organizations. Management training consultancies may also be able to provide this type of training. Effective presentation courses may be a good foundation. Liaison with Personnel department recommended. Reference books: Jay (1993)
10. Documentation techniques for HACCP Plans	HACCP Management Software (see listing in References, further reading and resource material) Word-processing skills training

The HACCP Team Leader will need a more advanced level of knowledge. This may be available through taught courses, but more likely it will be gained through an experiential approach, i.e. working with an experienced mentor on the application of the HACCP Principles, perhaps within your own factory.

Table 3.2 gives details of additional areas where training or knowledge may be required to support HACCP activities, and provides suggestions on how these gaps may be filled. It may not be necessary for all HACCP Team members to be trained in every area, but it will be helpful to have knowledge within the team.

3.2 What is our current status? – baseline audit and gap analysis

It is important to evaluate the resources and systems in place and compare these against the requirements to manage HACCP effectively, before putting together a Project Plan for the HACCP initiative. This will include an audit of your facility environment as well as an assessment of the current systems and personnel resources.

In order to plan the pathway to an effective HACCP System, it is important to consider two questions:

1. What resources and systems need to be in place for HACCP to work?
2. What resources and systems do I currently have?

The differences between 1 and 2 are the gaps that will need to be filled. A third question (How will I get there?) will be considered in section 3.3.

The most effective way of identifying the gaps is to carry out a baseline audit of current control measures for food safety and quality management, using audit skills and expert knowledge (possibly external to the company) of the standards and systems required to support HACCP.

3.2.1 Prerequisites and the HACCP Support Network

HACCP is a foundation for product safety management and, in practice, links with many other management systems, as illustrated in Figure 3.4.

In a manufacturing operation, it is unlikely that a HACCP System would be implemented as effectively in the absence of these other management systems. In a very small business, e.g. takeaway sandwich bar, these additional systems would probably be fewer, and might include Good Hygiene Practice and use of reputable suppliers.

Many of these systems could be called pre-requirements or 'prerequisites' for HACCP, as they are already known in some countries.

Prerequisite is the term used to describe systems that must be in place in order to support the HACCP System. The name *prerequisite* applies because these are normally systems in place *before* the HACCP Plan is developed. Whether you call these support systems, pre-requirements or prerequisites is of no consequence, but their role is essential for food safety control. We will use the term prerequisites throughout the rest of this book.

Prerequisites are normally systems in their own right, e.g. Supplier Quality Assurance, Good Manufacturing Practice. In fact, most of the systems in the HACCP Support Network, shown in Figure 3.5, could be called prerequisites, as they are all required to some extent for HACCP to function effectively. Prerequisite programmes are being used formally as part of HACCP System development in some countries, e.g. USA and Canada.

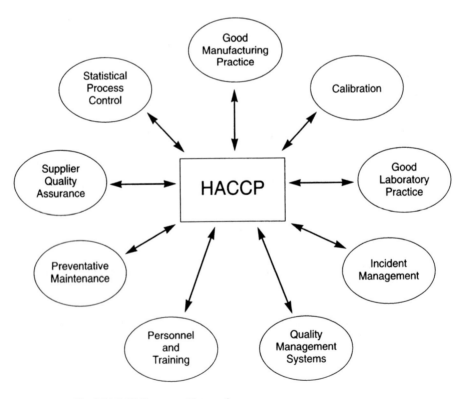

Figure 3.5 The HACCP Support Network.

HACCP and prerequisites – which comes first?

Like the riddle of the chicken and the egg, there is some argument about the order of introduction for HACCP and prerequisite systems into a company's quality management strategy. In some cases prerequisites might be better named post-requisites, as the HACCP process can be used to identify where the additional systems are essential.

However, if the prerequisite systems have been well designed and are in working order *before* you commence the HACCP process, then there may be less hazards to contend with in your operation than would otherwise potentially be present.

Prerequisites control the general day-to-day issues that, as responsible food processors, we ought to be controlling, e.g. not wearing jewellery in a food manufacturing area as part of a Good Manufacturing Practice programme. The types of 'hazards' controlled by these programmes tend to be those that, in reality, are unlikely to cause harm to the consumer, though they may be unpleasant, e.g. finding an earring in your yoghurt. We have to be careful, though, as in some countries, e.g. the USA, prerequisite programmes can include some legislative issues, such as thermal process requirements. The situation is further complicated by the fact that the term 'Good Manufacturing Practice' (GMP) does not have universal meaning. In some instances GMP is defined as including all food industry operating procedures (including HACCP and Quality Management System procedures), in others it focuses more on the building structural requirements, personnel and hygiene controls and sanitation.

Whatever the definition, well-planned prerequisite programmes will effectively design out generalized hazards that apply to the whole operation, leaving HACCP to deal with the specific significant product-process hazards. In the above example, where a prerequisite good manufacturing practice programme includes effective prevention of jewellery wearing by staff and visitors, you should not need to identify jewellery items as potential foreign material hazards during the HACCP Study. Basically, the prerequisite programme reduces the day-to-day likelihood of the hazard occurring.

Prerequisite programmes – the importance of verification

At the time of writing, there is still lively debate about the use of prerequisites. Some companies feel that prerequisite programmes are the way forward, allowing 'slimmed down', more focused HACCP Plans that deal with the significant product-process hazards. Other companies are wary of this approach, and consider that it is safer to manage all 'hazards' within the HACCP Plan.

Whichever approach is taken, there is no argument about one key issue – the need for verification. We will discuss verification of the HACCP System in detail in Chapter 8, as part of the system maintenance procedures. This is how we will demonstrate that the HACCP System is effective on an ongoing basis.

It is equally, if not more, important to verify prerequisite programmes, to ensure that they really are managing out environmental issues which are not then considered in the HACCP Plan. This is an issue that is easily forgotten when a company puts all its effort into development and implementation of the HACCP Plan. In this case, the HACCP System will only be effective as long as the prerequisites are also working effectively, so any verification of the HACCP System will be subject to additional verification of the prerequisite systems.

Prerequisite programmes can be an effective way to manage the repetitive hazards that occur throughout the facility at a number of possible locations. However, the dangers of then forgetting about these hazards, and simply concentrating on the specific hazards managed within the HACCP Plan, should not be overlooked.

3.2.2 Prerequisite support systems – the essentials

In this section we will consider the essential requirements for Good Manufacturing Practice and Supplier Quality Assurance prerequisite programmes. In Chapter 9 we will look at the other systems and will consider how to link them practically with HACCP. If you already have well-developed GMP and SQA programmes, you may wish to proceed to section 3.2.3.

(a) Good Manufacturing Practice

A key issue for product safety is the risk of cross-contamination occurring during the process from the internal factory environment. Cross-contamination could arise from a wide range of sources and the inherent risks in a particular processing area must be understood. Most of these issues are managed through adherence to Good Manufacturing Practice (GMP). Some of the main sources of potential cross-contamination are as follows:

Layout
The facility layout should be considered carefully to minimize the cross-contamination risks. This should include adequate segregation of raw materials and finished products. Depending on the type

of operation, full segregation between raw and cooked product may be required, and in most facilities the outer packaging stages, both for raw materials and finished products, will need to be kept separate from the main processing area. If you do not have the standards you require already in place, then layout upgrade and/or segregation can be timetabled into your Project Plan for HACCP development and implementation.

Availability of the required services and facilities for manufacture of the product should also be considered. This will include the availability of potable water, and adequate cleaning facilities for plant, equipment and environment, along with the connection of all required services in the correct area, e.g. steam heating and cooling facilities.

The number of holding stages and associated times should also be considered at this stage as it is important that there is adequate space for holding the required amount of product at each stage without causing a cross-contamination risk, and that the appropriate temperature-controlled facilities are available.

The patterns of movement of staff and equipment should also be assessed here, with the provision of adequate hygiene facilities, such as changing and rest rooms and handwash stations, along with canteen and recreational facilities.

Buildings

The fabric of the building itself could pose a hazard or safety risk to the product, through harbourage of pests and other contamination, or through physical contamination due to poor design and maintenance. Surfaces should be non-porous and easy to keep clean, with all cracks filled and sealed, and overhead services should be kept to a minimum. All buildings should be well maintained to prevent physical hazards falling into the product, and drains should be designed and serviced so that the flow is always away from production areas, with no chance of back flow or seepage. Adequate pest proofing and cleaning schedules should be drawn up for all facility buildings. All food manufacturing areas should be constructed such that these issues are managed.

Equipment

Equipment should be designed to minimize any cross-contamination risk. This could arise through parts of the equipment breaking off and gaining entry to the product as physical hazards. Alternatively, if equipment has any dead-leg areas, is difficult to clean or is poorly cleaned, microbiological build-up could contaminate the product. Chemical contamination could arise through lubricants or cleaning residues remaining on the equipment food-contact surfaces. Remember also to ensure that you can clean around and under equipment. If it is too close to the floor to clean underneath, the equipment should be sealed around the base.

You will also need to consider what the equipment is made of. For example, is it stainless steel or is it mild steel which may corrode leaving a surface prone to microbiological contamination? Is it painted and could your product be at risk from paint flakes? Does it have any wooden parts or brush attachments that cannot be effectively cleaned?

People

Food handlers and other personnel with access to the food processing area could cross-contaminate the product with microbiological, chemical or physical hazards. The process layout and movement patterns should be considered in order to minimize this risk, along with the appropriate training programmes.

Here also you will need to look at the types of protective clothing required, along with frequencies of changing and laundering procedures. You should already have considered changing facilities, amenities and handwash stations as part of the building layout, but cross-check whether you have made sufficient provision. All personnel in a food plant should be trained in Good Hygiene Practice. Steps should also be taken to ensure that this knowledge is fully utilized.

Cleaning

There must be sufficient facilities for the cleaning of equipment, people, plant and buildings, and these should be situated to enable their convenient use. Cleaning areas should not cause a cross-contamination risk to the process. Cleaning schedules should be prepared for all areas and staff must be adequately trained to carry out cleaning activities effectively.

Chemicals

Storage facilities must be provided for any chemicals that are required for use in the manufacturing area. These must prevent the risk of product contamination. All chemicals must be properly labelled and must not be decanted into food containers. All personnel handling chemicals must be trained in their safe use.

Raw materials

Raw materials can act as cross-contaminants if they gain access to the wrong product, or if they are added in excess quantities. This can have serious consequences in the case of allergenic raw materials entering a product where they are not labelled. Handling areas for raw materials must be carefully planned, and areas used for more than one type of ingredient may require thorough cleaning between use.

Make sure that you know how all your raw materials need to be handled, and put appropriate measures in place. You may be buying in something in a safe condition but it is easy to make it unsafe through improper handling, e.g. leaving perishable goods sitting on a loading bay for several hours.

Storage

Storage areas must be properly planned to minimize damage and cross-contamination issues. Consider whether you have adequate segregation, temperature and humidity control, and ensure that all storage areas are properly pest proofed. All materials should be stored off the floor and in sealed bags or containers. Part-used containers must be resealed after each use, and strict stock rotation should be employed.

Products

Residues of other products can also cause a serious hazard if allergenic material is present or if they affect the intrinsic nature of the product that is contaminated. Production lines should be spatially separated to prevent cross-contamination, and handling and cleaning procedures should be planned appropriately. Consider any extra control required if personnel are switched between lines or departments; will there be an additional risk from their protective clothing?

Packaging

Packaging areas and handling practices should be managed and controlled to prevent any cross-contamination risk. The packaging itself could be a major hazard, e.g. glass fragments, or could introduce micro-organisms to the product. Make sure that your packaging is suitable for the job and won't be damaged during product storage and distribution, and consider whether you have the correct coding and usage instructions printed legibly.

(b) Supplier Quality Assurance

One of the key areas for initial focus alongside HACCP development is raw material safety. We must understand the hazards associated with our raw materials if we are going to make a safe product. It is particularly important to know that the supplier is controlling hazards if these cannot be controlled in our process or by consumer action. For these reasons, an effective Supplier Quality Assurance (SQA) system is one of the most important prerequisite programmes. In this section we will consider how to build an effective SQA system.

There are a number of different elements to an effective SQA programme, including having agreed specifications, auditing suppliers and certificates of analysis. Supplier approval will depend on having confidence in the supplier's operation; that the supplier is competent at managing the hazards present. It is therefore vital to develop good customer/supplier relationships – partners in the management of safe raw materials and products.

Specifications

It is vital that all raw materials are purchased from approved suppliers to an accurate and up-to-date agreed specification. The specification is the cornerstone of your SQA system, detailing all the accepted criteria against which raw material quality and safety are measured. It should define clearly all the factors that you consider important to the raw material, and should include limits or tolerance of acceptability/unacceptability. The document can be as lengthy or as concise as you wish, but should always include your minimum acceptance criteria.

A typical raw material specification would include the following (these issues will also need to be addressed if you are buying in a finished product):

- Details of supplier and manufacturing/supply site.
- A description of the raw material and its functionality.
- An ingredients breakdown.
- Details of all intrinsic factors with tolerance limits, e.g. a_w, pH, salt, alcohol, etc.
- Microbiological acceptance criteria, e.g. absence of identified hazard organisms.
- Analytical and microbiological limits and sampling plans.
- Labelling requirements.
- Storage and distribution conditions.
- Safe handling and use instructions.
- Description of pack type, size and quantity.

A description of how the raw material is processed, or Process Flow Diagram, and a site plan are helpful to the HACCP Team in ensuring that they have fully identified all hazards of concern in the raw material. These are often supplied as separate documents, but this information will be essential when evaluating high-risk raw materials and should therefore be built into the specification. These documents can also be used to draw up a checklist of questions before the supplier audit. If your supplier is unwilling to provide processing information, perhaps for reasons of confidentiality, then you must be able to assure yourself that the raw material is safe by some other means. This may be through an understanding of the raw material's critical intrinsic factors along with the structured audit of the supplier's operation.

Auditing

Auditing is one of the key functions in any SQA system, as it is through audits that confidence can be gained in the supplier's operation. Before auditing a supplier there are a number of questions you will want to ask. Figure 3.6 provides an example of the

---

1. Company name, address, contacts and ownership details, including organizational structure and number of personnel.

2. Production site for this product.

3. How long has the factory been in operation?

4. Was the building purpose built?

5. Are any other types of product manufactured at this facility (and are any known allergens present on the site)?

6. Does the company operate a food safety management system based on the principles of HACCP?

7. Does the manufacturing site operate to a formal quality system such as ISO 9000, and is it certified?

8. Is microbiological testing carried out on site, and if so does this include pathogen testing?

9. Are any external contract laboratories used?

10. Have on-site and contract laboratories been accredited to an independent laboratory quality standard?

11. Is the manufacturing site covered by a pest control contract and, if not, what pest control procedures are in place?

12. Where is protective clothing laundered? If a contract laundry is used, has it been audited?

13. Who is responsible for plant hygiene? If contract cleaners are used, how often do they visit?

14. Are any raw materials, intermediate or finished products stored off-site, and if so who is responsible for the condition of these facilities?

15. Are specifications held for all raw materials and finished products?

16. Are written work procedures available on site?

17. Are there written personal hygiene standards?

18. What training do food handlers receive?

19. What vehicles are used for distribution (own/contract), and who monitors their condition?

20. What legislation is considered applicable to the company's operations?

Figure 3.6 Supplier Quality Assurance pre-audit survey.

type of pre-audit survey that might be sent to suppliers, but this list is by no means exhaustive. This information will also be important for low-risk raw materials when you do not intend to audit the supplier.

When you have constructed a programme of auditing requirements, it is important to think about how audits will be carried out. Do you, for example, have personnel who can carry out audits and are they trained appropriately? The SQA audit is important to the safety and quality of your products so it is vital that it is carried out effectively, and that you maintain good relationships with your suppliers. This can only be secured through choosing the correct type of personnel and training them properly.

We will discuss auditing in much more detail in Chapter 6, where we will be looking at auditing the HACCP System. The elements are the same for successful supplier auditing, except here we are looking more broadly at the supplier's entire operation.

Certificates of analysis
Certificates of analysis can be obtained for batches of raw materials to confirm that these have been sampled for certain criteria and providing the analytical result. You will need to check that they comply with specifications for these criteria. These certificates can form a useful part of the SQA system, but the limitations of end-

Figure 3.7 'Establishing a safe raw material supply'.

product inspection and testing (Chapter 2) should be remembered, and they should not be the only way of verifying that the finished product is free from the hazard(s).

You should ensure that certificates of analysis are prepared only by laboratories who are competent to carry out the tests and provide accurate results. This is best attained through following good laboratory practice and/or independent laboratory accreditation.

Third-party inspectors

If you do not have trained and experienced staff available to carry out your planned programme of audits, then you may wish to use third-party inspectors.

In choosing third-party inspectors, you will need to consider the expertise and experience of the auditors at the third-party inspection body. It is vital that the inspectors have sufficient experience both in the technology concerned and in auditing practices. You must be confident that they will highlight any potential food safety problems and help you to maintain good relationships with your suppliers. This can be achieved by going out and accompanying your suppliers' auditors to confirm that you are happy with their performance.

It is important to select a reputable inspection organization, and it is worth checking whether their inspections have been accredited by a higher-level board, e.g. against the requirements of EN 45004 (general criteria for the operation of various types of bodies performing inspection).

Buying from agents and brokers

When you buy raw materials through agents or brokers, you lose out on direct contact with the supplier. This can have drawbacks when the agent has little or no technical knowledge of the raw material, but it can work if you manage the situation effectively.

You must know how your raw materials have been processed and handled at every stage, in order to establish whether the likely hazards are present at expected or increased levels, and also whether any new unexpected hazards have crept in. It is important that you can obtain the appropriate assurances from the agent, and possibly from the supplier via the agent, and you must ensure that appropriate control is built into your own operation to cope with the worst-case scenario.

Even with the best-planned SQA system it is difficult to be absolutely sure that your raw materials always meet the required standards for safety and quality. In order to do this more effectively it is advisable to pass on to your suppliers the requirement to operate an effective

HACCP System for food safety hazards. This requirement can be passed right up the supply chain, so that at each stage – growers, processors, distributors, agents and final manufacturers – there is the same level of confidence in the material at that stage in the chain, in the same way as the consumer can have confidence in a finished product manufactured through an effective HACCP System.

3.2.3 Performing the Gap Analysis – questions to consider

A number of questions to consider when assessing the effectiveness of existing systems are:

1. Supplier Quality Assurance:

- Have all raw material suppliers been approved?
- Do you understand the safety criteria governing your raw materials?
- Does the raw material need to be handled in a specific manner?
- Have suppliers been audited? (And against which criteria?)
- What training do supplier approval auditors receive?
- Do suppliers provide analytical information and is it valid?
- Do you have confidence in the safety of all raw materials?
- Are specifications held for all raw materials?
- Are third-party audits carried out?

Consider also section 3.2.2(b).

2. Good Manufacturing Practice:

- Are your buildings and equipment in good repair?
- Is process flow logical
- Do you have effective cleaning programmes?
- Do you have procedures for the control of:

  jewellery,
  protective clothing,
  hand washing,
  personnel,
  hygiene behaviour,
  glass breakages,
  foreign material control?

- Are these procedures working effectively?

(Consider also section 3.2.2(a).

3. Personnel and Training:

- Do you have sufficient base line expertise?

- Are all personnel trained in food hygiene?
- Is additional training in HACCP required?

4. Good Laboratory Practice:

- Does the laboratory operate to a good practice system?
- Is the system accredited?
- Are controls built into the sample testing procedures?
- Is analyst performance monitored?
- Are all laboratory staff trained?
- Have external laboratories been approved?
- Is sampling carried out in a hygienic manner?

5. Quality Management Systems:

- Is there a Quality Management System in place?
- Is it based on an accepted framework, e.g. ISO 9001?
- Is it externally assessed?
- Does it cover all parts of the operation?

6. Preventative Maintenance:

- Does a preventative maintenance schedule exist?
- Does it cover all key equipment?

7. Recall and Incident Management:

- Does a system exist to manage serious incidents?
- Have recall procedures been tested?
- Would traceability systems ensure that all of the correct material could be withdrawn/recalled?
- Are personnel trained in media handling?

8. Statistical Process Control:

- Is process capability understood?
- Is SPC used in process monitoring and adjustment?

9. Calibration

- Are key pieces of process and monitoring equipment calibrated?

Following a gap analysis of the HACCP Support Network (i.e. the prerequisites and resources that need to be in place for HACCP to work), the HACCP application and implementation Project Plan can be developed by using standard project planning techniques (section 3.3.2).

3.3 How do we get there?

The next step is to plan the entire project, including the structure of the HACCP System. The project plan should include the progression of any necessary prerequisites (support systems) which were identified as gaps during the baseline audit.

3.3.1 What structure should the HACCP System take?

One of the key issues to decide early on is the structure of the HACCP System. This will depend on the complexity of the operation and types of processes being carried out, along with the status of Quality Management Systems already in place. There are three basic approaches:

(a) Linear HACCP Plans

In this approach the HACCP Principles are applied to each product or process on an individual basis, starting with the raw materials coming in and ending with the finished product. Depending on the type of operation, the HACCP Plans may be extended to include distribution and customer/consumer issues.

Linear or individual product/process HACCP Plans work best in simple operations, where there may be relatively few product types manufactured by a small number of processes. This approach is less likely to be helpful in larger, more complex operations. Here the application of HACCP Principles to each product/process becomes repetitive and needlessly time-consuming, and leads to a large number of similar HACCP Plans, each with its own management requirements.

(b) Modular HACCP Plans

If your products are manufactured using a number of basic process operations, it may be possible to use the modular approach when putting together a HACCP Plan. This flexible approach allows the HACCP Principles to be applied separately to each of the basic operations or modules. These HACCP Plan modules are then added together to make up the complete HACCP System.

For example, in a facility producing a number of ready meal products, several basic process operations may be in place, and each individual product will involve a combination of modules (Table 3.3). It is important to know where each module starts and ends so that no process step, and therefore no hazard, is missed when these are put together. Transfer steps from one module to the next can easily be missed and so should be clearly marked.

Since each module is specific to a part of the process and common to a number of products, the key issue is to ensure that the differences between products are picked up and all hazards addressed. Raw materials need to be assessed individually, considering their

Table 3.3 Process modules example – ready meal factory

Module	Process operation module

1	Storage – raw materials and packaging
	   chilled
	   frozen
	   ambient

2	Ingredient preparation
	   de-boxing
	   de-bagging
	   can opening
	   vegetable chopping/dipping/blanching
	   defrosting
	   weighing
	   mixing
	   pastry making

3	Manufacturing
	   mixing
	   baking
	   meat cooking
	   sauce cooking
	   pasta/potato/rice cooking
	   deep frying
	   assembly
	   mechanical/hand filling

4	Packing
	   mechanical/hand
	   coding
	   sleeving
	   outer casing

5	Storage and despatch
	   chilled
	   frozen
	   ambient

intrinsic hazards as they arrive and each use to which they will be put. Any special handling or processing measures used in a module for some products but not for others will also need to be considered.

The modular approach is a very effective way of structuring the HACCP System and is commonly used in complex manufacturing operations, and in catering operations which also split logically into a number of modular parts.

(c) Generic HACCP Plans

Generic HACCP Plans are based on a framework approach that is intended to fit similar operations where the same product is manufactured or handled. This approach has limitations because no two operations are exactly the same, and HACCP is designed to be applied to specific processes.

The danger with using purely generic HACCP Plans is that the issues that are specific to an operation may be overlooked, and therefore hazards may be missed out. Nevertheless, generic HACCP Plans can be used as a helpful starting point, and an effective HACCP System can be built up around them, ensuring that plant-specific hazards are managed in additional to generic hazards.

This type of approach is most commonly used in process sectors involving relatively simple operations; for example, primary meat processing. Bearing in mind the limitations discussed above, a generic HACCP Plan can be drawn up within a company for application to several sites, or generic plans published in the scientific literature can be adapted (e.g. AgriCanada, 1994; USDA, 1996b; Lee and Hathaway, 1998).

Using software to structure your system

A number of software packages are available that can help with the structure and documentation of your HACCP System. It is possible to generate satisfactory documentation for HACCP using standard word-processing skills. However, a number of specific HACCP packages are available, some of which may also be helpful as training tools.

A list of possible packages is given in the section on References, further reading and resource material. You should evaluate the capabilities of any package before planning to use it in structuring your HACCP System.

3.3.2 Using project planning techniques

When we use the word 'plan' in this section, we are referring to the development of a Project Plan and action timetable for the application and implementation of a HACCP Project, as opposed to the HACCP Plan, which will be covered in detail in Chapter 6.

The application and implementation of HACCP and supporting systems can be best managed as a project. It will have a definite life cycle with a start date and a finish date when HACCP can be said to be fully operational. In a larger organization, the project may be managed by a temporary project team and the timetable and costs estimated at the start. This will involve the appointment of a few key people and the documentation of the actions and time scale required. The roles needed in managing the project are two key personnel plus a supporting team.

The Project Sponsor

As the champion of the project, the Project Sponsor is likely to be your company Managing Director, Operations Director or Technical Director. Whoever takes on the role is likely to sit on the senior management team and have budgetary control. The main responsibilities are to:

- provide funds;
- approve and drive the company HACCP or food safety policy;
- approve the business issues and ensure that the project continues to move forward and remains valid;
- appoint a Project Manager and Team;
- ensure that adequate resources are made available to the Project Team;
- establish a progress reporting procedure;
- ensure that the Project Plan is realistic and achievable;
- approve any changes to the original project.

The Project Manager

This role is likely to be taken by the Production or Technical Manager, who may also go on to become the HACCP Team Leader. The responsibility centres on ensuring that the Project Plan is drawn up and objectives achieved within the agreed time scale. This requires effective project management skills, specifically to:

- lead and direct the Project Team;
- produce an achievable Project Plan;
- provide a regular progress report to the Project Sponsor;
- liaise with other Project Managers to ensure that areas of common interest are identified and resources are used effectively in these areas.

As well as the HACCP Project itself, there may well need to be other business improvement projects going on within the business at the same time. These may include development of the systems required to fill the gaps in the HACCP Support Network, which we identified following the baseline audit (section 3.2). It is useful to establish what these additional systems are early on, for example as in Figure 3.8.

Other projects may include the setting up of a formal Supplier Quality Assurance programme, the introduction of Statistical Process Control on certain lines, production rationalization, facility environment alterations, product development activities, cost of quality projects, and so on.

The project team will need a complete understanding of the

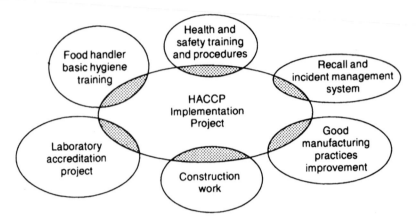

Figure 3.8 HACCP interaction with business improvement projects.

starting point. This will be most simply achieved if all the project team members have already been involved in the baseline audit, including a review of documented procedures already in place, environmental issues, GMP status, resource availability and current culture. In other words, the team will have an appreciation of the size of the task, the current capabilities and the additional resource requirements.

3.3.3 Drawing up the Project Plan

Once you have decided on the stru cture of the HACCP System and have an understanding of the additional tasks you will need to undertake, a Project Plan can be drawn up. The complexity of this plan will relate to the amount of work to be done and it is important to ensure that sufficient time is allocated to develop an effective system.

The Project Plan may be divided into a series of main phases, with each phase further broken down into specific activities – a work breakdown structure. The start, finish and activity times are determined, together with any dependencies (What needs to happen before this particular activity can take place?) and the resource allocated (Who will make it happen?). This can be plotted on paper to provide an implementation timetable known as a Gantt chart. An example is given in Figure 3.8. In looking at the Gantt chart, notice that while the duration of each task has been esti-mated, not all tasks can begin on day 1. This is because some of them cannot start until another task has been completed.

The following are definitions of key terms introduced on the Gantt chart:

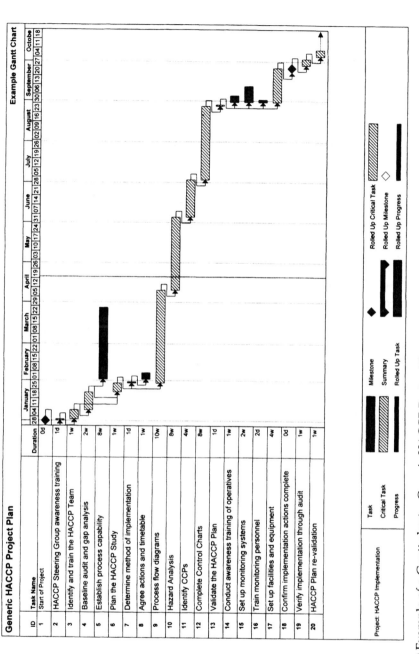

Figure 3.9 Example of a Gantt chart – Generic HACCP Project Plan.

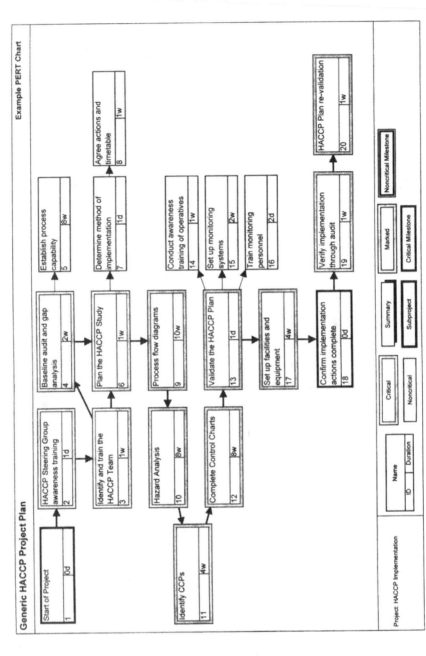

Figure 3.10 Example of a PERT chart – Generic HACCP Project Plan.

- Critical: this means that the task is critical in terms of timing. If these tasks do not run to time then the project completion date will be affected – there is no slack.
- Non-critical: this does not mean that the tasks are any less important than those referred to as Critical. It just means that there is some slack in the timing. If they don't finish at the precise date indicated, then depending on how long over they run, the end project completion date may not be affected.
- Milestone: this is usually an event, or key decision date. It can be used as an indicator in terms of the project progress.

By then considering the start and finish times and the dependencies as shown on the Gantt chart, a dependency network can be established. This is also known as a PERT chart (Programme Evaluation and Review Technique). A PERT chart is given in Figure 3.10 for the example of the Generic HACCP Project Plan.

The PERT chart indicates the Critical Path (in bold lines) for the project. This is the shortest possible time in which the project could be achieved. It should be noted that, to be truly useful, the task duration must be calculated using the elapsed time taken to complete a task and not the number of man days needed. As an example, the hazard analysis has been set at 8 weeks. This could be made up, for example, of 16×3-hour HACCP Team meetings held during that period.

Both Gantt and PERT charts can be produced using computer software packages. They can also be drawn manually and kept fairly simple, e.g. this may be appropriate for a smaller business. The examples shown here were developed using Microsoft Project™.

3.4 Summary

As we saw in Figure 3.1, a successful HACCP System results from following four key stages. In this chapter we have discussed key stage one, Planning and Preparation (Figure 3.11). Of course to plan any system effectively, the scope of the entire project needs to be understood at the beginning. As previously stated, it is therefore important to read the rest of the book before coming back to developing the Project Plan. Stages two, three and four will be covered in detail in Chapters 6, 7 and 8, respectively.

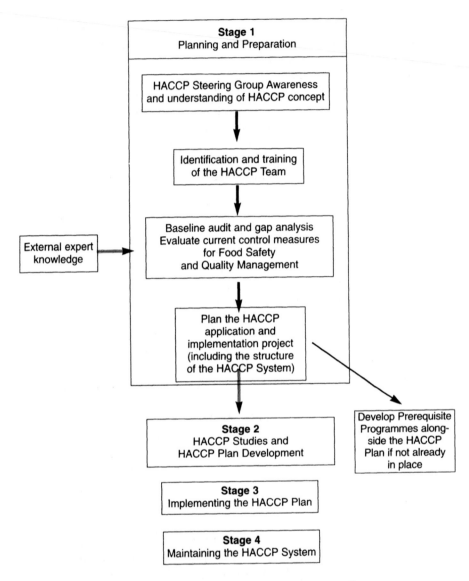

Figure 3.11 Key stage 1 – Planning and Preparation summary diagram.

4

An introduction to hazards, their significance and control

This chapter is designed to give you a clearer understanding of different types of hazards and their significance in foods. We will explore the mechanisms that can be used for their control and expand on the concept of Critical Control Points. This information is intended for reading before you get going and, although not a complete source of information on all hazards, it will give a valuable grounding to the HACCP Team members. You may not initially have people who will understand the implications of all the information given here, but it will help you to get started and will highlight areas where you may need to bring in specialist help to your HACCP Team.

4.1 Hazards and their significance

A hazard is any factor that may be present in the product, which can cause harm to the consumer either through injury or illness. Hazards may be biological, chemical or physical and are the basis of every HACCP System.

HAZARD:
A biological, chemical or physical property, or condition of, food with the potential to cause an adverse health effect.

(Codex 1997)

Physical hazards are the most common type of hazard to occur in foods, because of the possible presence of foreign material. However, the risk of consumer injury is quite low for most types of foreign material, as few items are sharp or hard enough to cause physical damage, or are of dimensions that might cause choking. When present, physical hazards may affect only one or, at most, a few people.

Chemical hazards, e.g. pesticides and hormones, are often looked on as the most important by the consumer but in reality they often pose a negligible health risk at levels likely to be found in food. On the other hand, biological hazards usually present the greatest and broadest danger to consumers. When a pathogenic micro-organism grows in a food product, it can cause illness in many hundreds or thousands of consumers. Some of these illnesses can be quite serious, even fatal.

Looking at the factors contributing to food poisoning outbreaks in England and Wales during 1970–1982 (Table 4.1; Hobbs and Roberts 1993), we can see that better knowledge of how to control the likely hazards may have prevented many of the examples given. Although these data were collected some time ago, there is little evidence to suggest that there has been a major change.

In order to assess which identified hazards must be controlled by the HACCP System, their significance needs to be evaluated at the hazard analysis stage (Chapter 6).

This involves considering each potential hazard in turn and attempting to answer the questions: 'Could this hazard reasonably be expected to occur in my raw materials, process or product?' and 'Would it cause harm to the consumer?' This needs to be an educated judgement taken by appropriate, experienced personnel based on the best information. If incorrect judgements are made at this stage, the resulting HACCP System will be unsound, and the company will be operating under a false sense of security. It is therefore essential to have the correct blend of expertise and information available, and inexperienced HACCP Teams must recognize their limitations and supplement these with extra help where required.

4.2 Hazard types

This section is intended as an introduction to different types of food hazards. It does not contain exhaustive lists of hazards, and should not be used as a replacement for the correct blend of knowledge and experience within the HACCP Team. Instead, the information provided should be taken as suggestions for possible further investigation.

4.2.1 Biological hazards

Most food-processing operations will be at risk from one or more biological hazards, either from the raw materials or during the

Table 4.1 Factors contributing to 1479 outbreaks of microbiological food poisoning, England and Wales, 1970–82

Contributing factors	Number of outbreaks in which factors recorded (%)					
	Salmonella	C. perfringens	Staph. aureus	B. cereus	Other	Total
Preparation too far in advance	240 (42)	464 (88)	80 (48)	54 (86)	6 (4)	844 (57)
Storage at ambient temperature	172 (30)	276 (53)	75 (45)	39 (62)	4 (3)	566 (38)
Inadequate cooling	125 (22)	313 (60)	12 (7)	17 (27)	1 (<1)	468 (30)
Inadequate reheating	76 (13)	275 (52)	5 (3)	33 (52)	2 (1)	391 (26)
Contaminated processed food	100 (19)	19 (4)	27 (16)	4 (6)	86 (54)	246 (17)
Undercooking	139 (25)	74 (14)	2 (1)	1 (2)	7 (4)	223 (15)
Contaminated canned food	2 (<1)	4 (<1)	42 (25)	1 (2)	55 (35)	104 (7)
Inadequate thawing	61 (11)	34 (6)				95 (6)
Cross-contamination	84 (15)	8 (2)	2 (1)			94 (6)
Raw food consumed	84 (15)		1 (<1)			93 (6)
Improper warm holding	15 (3)	52 (10)		8 (13)	8 (5)	77 (5)
Infected food handlers	13 (2)		50 (30)		2 (1)	65 (4)
Use of leftovers	25 (4)	25 (5)	11 (7)	1 (2)	2 (1)	62 (4)
Extra large quantities prepared	29 (5)	17 (3)	2 (1)			48 (3)
Total	566	525	166	63	159	1479

Adapted from Hobbs and Roberts (1993).

process, and the HACCP Plan will be designed to control these. Biological hazards can be either macrobiological or microbiological.

Macrobiological issues, such as the presence of flies or insects, while unpleasant if found, rarely pose a risk themselves to product safety in its true sense. There are a few exceptions to this, such as poisonous insects, but on the whole the appearance of macrobiological hazards simply causes revulsion. However, they may be an indirect risk by harbouring pathogenic micro-organisms and introducing these to the product. For example, an insect harbouring *Salmonella* could pose a major risk to the consumer if it gained access to a fresh, ready-to-eat product. However, the same insect gaining access to a canned product before retorting would not be a true food safety issue as it would be sterile in the finished product. Obviously it is important to ensure that your products are free from macrobiological hazards and you may want to consider them as part of your HACCP Study or prerequisite programmes. It is usual to consider macrobiological issues as foreign material or physical contaminants, rather than biological hazards.

Pathogenic or disease-causing micro-organisms exert their effect either directly or indirectly on humans. Direct effects result from an infection or invasion of body tissues and are caused by the organism itself, e.g. bacteria, viruses and parasites/protozoa. Indirect effects are caused by the formation of toxins (or poisons) that are usually pre-formed in the food, by bacteria and moulds.

Bacteria are broadly divided into two types, depending on a simple colour reaction produced by the Gram stain. Bacteria are therefore classified as either Gram-negative or Gram-positive. As a general rule, Gram-negative bacteria tend to exert their effects through invasion of the host (foodborne infection) whereas the effects of Gram-positive bacteria are usually mediated via pre-formed toxins (food poisoning).

Consequently the infections caused by Gram-negative bacteria generally have an onset period of 24 hours, are long lasting and debilitating. They are rarely fatal in healthy individuals but can cause death in the young, old, ill or immunocompromised, e.g. *Salmonella*. Illness caused by pre-formed toxins of Gram-positive bacteria have a rapid onset period of 1–6 hours, are short lived (lasting 24–48 hours) and are not usually fatal, e.g. *Staphylococcus aureus*. This is an oversimplification and, as with most biological systems, there are always exceptions. For example, Gram-positive *Clostridium botulinum* produces a lethal toxin, *Listeria monocytogenes* causes abortions and meningitis, and the effects of some Gram-negative bacteria, e.g. *Escherichia coli*, are mediated via toxins. Accordingly, the reader is strongly advised to seek expert professional advice.

(a) Pathogenic Gram-negative bacteria

The Gram-negative pathogenic bacteria typically associated with foods include *Salmonella, Shigella, Escherichia coli, Campylobacter jejuni, Vibrio parahaemolyticus, Vibrio vulnificus* and *Yersina enterocolitica*. These are usually present in the intestine and faeces of man, animals and birds. Consequently they can also be found in soil, water, raw agricultural products such as raw milk and raw meat (particularly poultry) and shellfish. These bacteria are not particularly heat resistant and will generally cause problems as a result of poor sanitation, inadequate personal hygiene and the cross-contamination from raw materials to work surfaces, utensils, processing equipment/machinery, finished products and packaging. Control is mediated by heat processing (e.g. pasteurization), segregation of raw and cooked foodstuffs, good hygienic working practices and/or formulating and storing the product such that the pathogen is inactivated or prevented from growing (e.g. fermented raw sausage).

There are about 2000 serovars of *Salmonella* spp., most of which are capable of causing foodborne illness in humans. Salmonellae grow in the intestines of all animals and are a common contaminant of raw meat, poultry, eggs and dairy products. Dried eggs and dairy products should be managed as *Salmonella*-sensitive ingredients. Capable of survival for long periods in frozen and dry conditions, salmonellae are 'ubiquitous' in the natural environment. They can be a persistent environmental contaminant in food plants.

A pandemic of *S. enteritidis* phase type 4 related to shell eggs emerged in 1980. To some extent the pandemic has been ameliorated by the widespread use of liquid pasteurized eggs instead of shell eggs in the food-service industry. *Salmonella typhi*, the cause of typhoid fever, is spread primarily by contaminated water.

Shigella dysenteriae produces several toxins that cause dysentery. Transmitted by the faecal–oral route, the organism does not survive well in processed foods. It is a public health threat primarily when infected food handlers work in the food-service industry.

Most strains of *Escherichia coli*, a universal intestinal inhabitant, are harmless to their human and animal hosts. However, several strains are capable of causing foodborne infections. The most serious of these are the enterohaemorrhagic (EHEC) strains, represented by *E. coli* O157:H7. All EHEC strains produce verotoxins, or shiga-like toxins, that produce bloody diarrhoea. Serious infections can proceed to haemolytic uraemic syndrome and renal failure. *E. coli* O157:H7 is the most common cause of renal failure in children.

First detected in foods in 1982, *E. coli* O157:H7 was found to inhabit some dairy cattle, from which contaminated minced meat and raw milk were found to be responsible for illness outbreaks. Since 1982 the host range of this organism has expanded to other animals. Unlike most foodborne pathogens, *E. coli* O157:H7 is very acid-tolerant. It has been found to survive and cause illness in fermented sausages, mayonnaise and unpasteurized fruit juices. Effective control of this organism depends principally on adequate cooking or pasteurization of foods. A serious outbreak of foodborne infection related to this organism occurred in Scotland in 1996, in which 20 elderly people died and 496 were infected in total (The Pennington Group, 1997).

Campylobacter jejuni is the most common cause of bacterial gastroenteritis in the UK and the USA. It is found principally in raw poultry and, unlike the other enteric pathogens, it does not grow well in foods as it requires exacting conditions for growth. The food itself is merely a vehicle (or vector) for infection, so segregation and inactivation via thermal processing are the most effective control measures.

Vibrio parahaemolyticus is more salt tolerant than the other Gram-negative pathogens and is found in marine environments and animals. This bacterium is typically associated with raw or under-processed seafood, and accounts for 50–70% of food poisoning in Japan.

Vibrio vulnificus, like the other species of *Vibrio*, is associated with seafood and the marine environment. The organism is highly invasive and causes primary septicaemia. Its virulence appears to be enhanced in individuals suffering from hepatitis or chronic cirrhosis, where it can be fatal. Other *Vibrio* species can also cause gastroenteritis, for example *V. cholerae*, which is associated with waterborne gastroenteritis.

Yersinia enterocolitica belongs to the same family as *E. coli* and *Salmonella*. It is a ubiquitous organism that has been associated with a wide variety of foodstuffs and, like *Listeria*, has the ability to grow at low temperatures. It can also produce an enterotoxin. The major sources of pathogenic types of *Yersinia* are raw pork, raw milk and water. Outbreaks have been associated with these sources and from pasteurized milk. Thermal processing and the prevention of post-process cross-contamination are the principal methods of control.

(b) Pathogenic Gram-positive bacteria

The Gram-positive pathogens are a diverse and unrelated group of organisms, including *Clostridium botulinum*, *Clostridium perfringens*, *Bacillus cereus*, *Staphylococcus aureus* and *Listeria monocytogenes*.

Species of the genus *Clostridium* are anaerobic, i.e. they grow in the absence of oxygen, and produce heat-resistant spores. They are generally widely distributed in nature and are usually found in soil, vegetation, fresh water and marine sediments and animal faeces. Consequently, their elimination and control is achieved by processing with high temperatures (such as those involved in canning) and product formulation, e.g. adding acids (pickling) or reducing the available water (preserving with sugar or salt).

Clostridium botulinum is important because it produces a lethal neurotoxin that paralyses the respiratory muscles. Historically, botulism has been associated with underprocessed canned foods, particularly home-canned foods. In recent years, cases of botulism have been caused by a wider range of foods, including improperly handled baked potatoes and garlic-in-oil preparations.

Strains of *Cl. botulinum* are broken down into groups on the basis of the toxin type that they produce. There are two sets of strains that must be considered by the HACCP Team. The first of these (usually in toxin groups A, B or F) produce quite heat-resistant spores and are highly proteolytic, thereby producing obvious signs of putrefaction when growing in a food. This proteolytic set does not grow below 10 °C. It is found in intestinal tracts and soil; therefore, it is a common contaminant of vegetables.

The second set (usually in toxin groups B, E, F or G) produces weakly heat-resistant spores and is non-proteolytic, usually producing no signs of spoilage when growing in food. These strains are psychrotrophic, capable of growth at 3.3 °C. They are commonly found in aquatic environments; the usual foodborne sources are fish and other seafood.

Botulism typically occurs when an individual consumes preformed toxin in a food. The botulinal toxins are heat-labile and readily inactivated by cooking. In rare cases botulinal spores from food or soil can be ingested and grow in the intestine if the usual microflora is not in place. Infant botulism occurs in this manner. It has sometimes been associated with the use of honey in infant formulae.

In comparison, the mode of action of *Cl. perfringens* is quite different. Food poisoning due to this organism is usually associated with undercooked and/or inadequately reheated meats and sauces, particularly in catering operations. It can grow rapidly at temperatures as high as 50 °C. The organism grows to large numbers in the food and produces its toxin during spore formation in the intestine after consumption. The toxin causes diarrhoea and nausea but is not normally fatal. The controls required are effective thermal processing, rapid chilling, segregation of raw and cooked materials, chilled storage of cooked meat before consumption, and adequate reheating and hot storage before consumption.

In contrast to the clostridia, species of the genus *Bacillus* are normally aerobic spore forming, i.e. they need oxygen to grow.

Bacillus cereus also produces two types of toxins: a very fast-acting emetic toxin which causes vomiting, and a diarrhoeal toxin. The former is very heat stable and will survive cooking; the latter is easily inactivated by cooking. The organism is commonly found in soil, vegetation and raw milk. Food poisoning is frequently associated with cooked rice and other starchy products, where the spores have not been inactivated by the initial heat process and have subsequently been allowed to germinate and grow due to inadequate handling and poor temperature control.

Unlike the other Gram-positive pathogenic bacteria, the sources of *Staphylococcus aureus* are frequently human in origin, i.e. from the skin, nose, throat, cuts and sores. Consequently it is easily transmitted to any foods by handling and poor hygienic practices. If the bacterium is allowed to grow in foods, it will produce a toxin that is stable to further heat processing and therefore cannot be made safe again.

Staphylococcus aureus does not form heat-resistant spores and is more versatile than other pathogens because it is able to tolerate a greater range of growth conditions. For example, it is capable of growth at water activity values as low as 0.86; however, it will not normally produce toxin below 0.92. Accordingly, strict personal hygiene is of paramount importance for the control of this organism, as well as thermal processing and segregation.

The importance of *Listeria monocytogenes* as an agent of foodborne disease has only recently become fully recognized. It is important because it has a high mortality rate in immunocompromised individuals. The foetus, for example, is particularly susceptible; spontaneous abortion is a frequent complication of listeriosis in pregnant individuals. *Listeria monocytogenes* is widely distributed in the soil, vegetation and animal faeces. It is, therefore, a common contaminant of raw foods. It is psychrotrophic, able to grow at 1 °C, and proliferates in cool, moist processing environments. Production controls are necessary to avoid environmental contamination of cooked foods during handling and packing.

The expanding array of psychrotrophic foodborne pathogens, *L. monocytogenes, Y. enterocolitica* and non-proteolytic *C. botulinum*, has drawn a great deal of attention to the safety of perishable refrigerated foods. It is highly recommended that such foods be restricted to a short shelf-life unless food safety barriers in addition to refrigeration are incorporated into the food.

(c) Emerging pathogens

The term 'emerging pathogens' is used to describe those organisms that have not historically been recognized as agents of human disease. Two such organisms are *Aeromonas hydrophila* and *Plesiomonas shigelloides*. They are closely related Gram-negative bacteria. Both are aquatic bacteria that can also be recovered from a variety of aquatic and terrestrial animals. Both are associated with diarrhoea in man, but whether either is a primary pathogen is not entirely clear.

They are inactivated by pasteurization and the controls required are thermal processing and the prevention of cross-contamination. However, these procedures may not be appropriate for raw seafood and shellfish, where the hazards must be recognized and accepted or other decontamination procedures established.

In the past 15 years several prominent pathogens have 'emerged' as serious threats to human health. These include *L. monocytogenes*, *E. coli* O157:H7, *Vibrio vulnificus* and *Y. enterocolitica*. This is an important lesson for food safety professionals who should expect the continued emergence of new foodborne microbial pathogens in the future. Continued diligence on the part of microbiologists, epidemiologists and HACCP Teams will be necessary to quickly identify and control new pathogens as they emerge.

(d) Viruses

Viral gastroenteritis is believed to be second only to the common cold in frequency and greatly exceeds the incidence of foodborne bacterial gastroenteritis. There are several types of viruses but the greatest number of outbreaks are due to hepatitis A and small round structured viruses (SRSV) such as the Norwalk virus. Shellfish (particularly molluscan shellfish) are the most common food source because they concentrate the virus from contaminated water. Despite this, much less is known about the incidence of viruses in food than about bacteria and fungi. This is because viruses are obligate parasites; they do not grow on culture media or in foods (food is a vector only). In addition they are very small and therefore very difficult to detect.

Viruses are present in man, animals, faeces, polluted waters and shellfish. They are transmitted from animals to people and from person to person. Hence high standards of personal hygiene are essential.

Viruses can sometimes be transmitted by foods contaminated by infected food handlers. A large outbreak of Norwalk virus was caused by contaminated pastry dough (Kuritsky *et al.*, 1984). Contaminated frozen strawberries caused an outbreak of hepatitis A (Calhoun County Department of Public Health *et al.*, 1997). Both viruses are readily inactivated by heat.

(e) Parasites and protozoa

The larvae of parasites such as pathogenic flatworms, tapeworms and flukes may infect man via the consumption of the flesh of infected pork, beef, fish and wild game. Examples include *Taenia saginata* (beef tapeworm), *Trichinella spiralis* (nematode in pork) and *Clonorchis sinensis* (trematode or fluke from Asian fish). Prevention of parasite infestation is achieved by good animal husbandry and veterinary inspection, along with heating, freezing, drying and/or salting, the most effective methods being heating (>76 °C) and freezing (−18 °C).

Protozoa such as *Toxoplasma gondii, Giardia intestinalis (lamblia), Cyclospora cayetanensis* and *Cryptosporidium parvum* produce encysted larvae which subsequently infect man on ingestion. Infected meat and raw milk serve as the sources for *Toxoplasma*, whereas raw milk and drinking water are the sources of *Giardia, Cyclospora* and *Cryptosporidium*. Human infection can also be contracted by direct contact with infected pets and animals. The oocysts are resistant to chemical disinfection, but can be inactivated by heating, freezing and drying.

While most causes of protozoan parasites are waterborne infections, contaminated foods are a possible vector. A large outbreak of *Cyclospora* was traced to fresh raspberries which had been sprayed with a pesticide mixture that had been prepared with contaminated water (Herwaldt and Ackers, 1997).

(f) Mycotoxins

Mycotoxins are produced as secondary metabolites of certain moulds, and can cause long-term carcinogenic effects at ongoing low levels, or short-term acute toxic effects at high levels of exposure. Although a large number of mycotoxins have been isolated, only a very small minority have been shown to be toxic to mammals. Mycotoxins are being considered here under biological hazards because they are products of microbial growth, although they may also be considered as chemical hazards.

The mycotoxins of concern in food production are aflatoxins, patulin, vomitoxin and fumonisins, which may be consumed by humans via two routes: firstly, direct consumption of contaminated grain or, secondly, indirect consumption via animal products, such as milk and turkey meat.

Aflatoxins

The most important group of foodborne mycotoxins, these are normally controlled through legislation at the commodity level.

Most countries have established regulatory limits on the presence of aflatoxin. These mycotoxins are produced by *Aspergillus flavus* and a few other moulds growing on foodstuffs. There are six aflatoxins of concern, four of which (B1, B2, G1 and G2) occur in various foods, and two of which (M1 and M2) are metabolites found in the milk of lactating animals that have eaten aflatoxin-contaminated feed. Aflatoxin B1 is found most commonly and occurs in groundnuts and in grain crops, particularly maize.

Aflatoxins normally contaminate crops during the growing or storage periods. During the growth period the aflatoxin contamination risk is increased by those environmental conditions which stress the plants and allow contamination with the mould, for example insect damage or drought conditions. Poor storage conditions, such as dampness and humidity, will increase the chance of unacceptable contamination. In order to control aflatoxins in your products you must understand the risks associated with each raw material source and with storage at your facility. Appropriate control can then be built in as part of your HACCP System.

Patulin

This is a mycotoxin associated with fruit and fruit-juice products. Produced by several *Penicillium* spp., it is considered to be a carcinogen, and high concentrations may cause acute effects such as haemorrhages and oedema. The presence of patulin in food products is normally associated with the use of mouldy raw materials, and this can be prevented by building effective control measures into your HACCP System.

Vomitoxin

Also known as deoxynivalenol (or DON), this is a member of the tricothecenes group, which is of particular importance and has been found in grain crops worldwide. It is produced by the mould *Fusarium graminearum* and its presence is promoted by wet weather conditions. This mycotoxin is known to cause toxic effects in animals, and human illness has been reported. It is controlled through legislation in some countries where there have been particular problems, and should be considered as a raw material issue in the HACCP Study, where grain crops are concerned.

Fumonisin

Produced by *Fusarium moniliformi*, fumonisin has been linked epidemiologically to oesophageal cancer in humans and has serious toxic effects in some animals, particularly horses. It is usually associated with maize and must be controlled at the commodity level. Regulatory limits are being considered in several countries.

Many other mycotoxins are produced by fungi, but none of these are currently considered to be of risk in the food supply. However, you must ensure that your HACCP Team has access to sufficient expert knowledge so that any newly discovered mycotoxin issues can be considered for your products.

4.2.2 Chemical hazards

Chemical contamination of foodstuffs can happen at any stage of their production, from growing of the raw materials through to consumption of the finished product. The effect of chemical contamination on the consumer can be long term (chronic), such as for carcinogenic or accumulative chemicals (e.g. mercury) which can build up in the body for many years, or it can be short term (acute), such as the effect of allergenic foods. The current main chemical hazard issues in food products are as follows.

(a) Cleaning chemicals

In any food preparation or production operation, cleaning chemicals are one of the most significant chemical hazards. Cleaning residues may remain on utensils or within pipework and equipment and be transferred directly onto foods, or they may be splashed onto food during the cleaning of adjacent items.

It is therefore vitally important that the HACCP Team members consider the implications of the cleaning procedures in their operation. Problems can be prevented by the use of non-toxic cleaning chemicals where possible, and through the design and management of appropriate cleaning procedures. This will include adequate training of staff and may involve post-cleaning equipment inspections.

(b) Pesticides

Pesticides are any chemicals that are applied to control or kill pests, and include the following:

- insecticides;
- herbicides;
- fungicides;
- wood preservatives;
- masonry biocides;
- bird and animal repellents;
- food storage protectors;
- rodenticides;

- marine anti-fouling paints;
- industrial/domestic hygiene products.

Pesticides are used in a wide range of applications all over the world – in agriculture, industry, shipping and the home. The use most relevant to food safety is in agriculture but contamination from other sources must also be considered.

In agriculture pesticides are used during production to protect crops and improve yields, and after harvest they are again used to protect the crops in storage. However, not all pesticides are safe for use in food production (for example, some of those used for the treatment of timber) and even those that are safe for food use may leave residues that could be harmful in high concentrations. To overcome these problems most countries have very strict control on the pesticides that can be used and on the residue limits that are acceptable. These are set through expert toxicological studies and are normally laid down in legislation.

From the food safety point of view you need to know which pesticides have been applied to all your raw materials at any stage in their preparation. You must also understand which pesticides are permitted for use and what the maximum safe residue limits are in each case. Control can be built into your HACCP System to ensure that the safe levels are never exceeded in your products.

In addition to raw materials that have direct pesticide contact, you must also consider the possibility of cross-contamination with pesticides at any stage in food production. This could be cross-contamination of your raw materials or it could happen on your site, e.g. from rodenticides. These issues should again be considered as part of your HACCP Study or prerequisite programmes. Additional information on pesticides can be found in CCFRA (1995a) and in national legislation.

(c) Allergens

Some food components can cause an allergic or food intolerance response in sensitive individuals. These reactions can range from mild to extremely serious, depending on the dose and the consumer's sensitivity to the specific component. Extreme anaphylactic responses are seen in individuals with severe allergies. Major allergens of concern include:

- peanuts (groundnuts);
- tree nuts (walnuts, hazelnuts, pecans, cashews);
- eggs;
- milk products;
- shellfish.

More minor allergens tend to include:

- soybeans;
- wheat gluten;
- certain colourings;
- sulphites;
- other tree nuts not listed above.

Allergens of concern may vary from country to country.

The control options open to the food processor manufacturing products with allergenic components are raw material control, effective pack labelling, control of rework and effective cleaning of equipment. The label must describe the product contents accurately, highlighting any potentially allergenic components. Special care must be taken when declaring a generic category of ingredient such as 'fish' or 'nuts', where certain individuals may be allergic to specific species of fish or type of nut. A manufacturer or caterer who produces several different products must also consider the chance of cross-contamination of allergenic components into the wrong product where they will not be labelled. This is particularly important in the case of recycling loops and rework of product, and these issues should be considered as part of the HACCP Study. The possibility of mislabelling through using misprinted or incorrect packaging, e.g. packing a ready meal product into the wrong sleeve, should also be evaluated.

Recent trends have seen food manufacturers using 'catch all' warnings on product packaging, for example 'Warning : this product may contain traces of peanut.' This is normally done where a number of products containing nuts are manufactured on the same line as non-nut containing products, e.g. in breakfast cereal manufacturing, or where rework is involved which may have been in contact with nuts, e.g. in the confectionery industry where enrobing chocolate is reclaimed. Such labelling is only felt to be helpful by anaphylaxis sufferers when no other control options are possible as it is otherwise seen as a limitation of their diet. Manufacturers are challenged to find better ways of preventing cross-contamination with these allergens.

(d) Toxic metals

Metals can enter food from a number of sources and can be of concern at high levels. The most significant sources of toxic metals to the food chain are:

- environmental pollution;
- the soil in which food stuffs are grown;

- equipment, utensils and containers for cooking, processing and storage;
- food-processing water;
- chemicals applied to agricultural land.

Particular metals of concern are tin (from tin containers), mercury in fish, cadmium and lead, both from environmental pollution. Also significant are arsenic, aluminium, copper, zinc, antimony and bismuth, and these have been the subject of research studies.

Just as for any other chemical hazard, you need to understand the particular risk of toxic metals to your products, and this is likely to be associated with the raw materials, metal equipment and finished-product packaging. Control can be built in as part of your HACCP System.

(e) Nitrites, nitrates and *N*-nitroso compounds

Nitrogen occurs naturally in the environment and is present in plant foodstuffs. It is also a constituent of many fertilizers, which has increased its presence in soil and water.

Historically, nitrites and nitrates have been added to a number of food products as constituents of their preservation systems. This deliberate addition of nitrite and nitrate to food is closely governed by legislation as high levels of nitrites, nitrates and *N*-nitroso compounds in food can produce a variety of toxic effects. Specific examples include infantile methaemoglobinaemia and carcinogenic effects.

N-nitroso compounds can be formed in foods from reactions between nitrites or nitrates and other compounds. They can also be formed *in vivo* under certain conditions when large amounts of nitrites or nitrates are present in the diet. In common with a number of other chemicals, nitrate can cause additional problems in canned products, where it can cause lacquer breakdown, allowing tin to leach into the product.

The HACCP Team must ensure that nitrite and nitrate being added to products do not exceed the legal, safe levels and must give appropriate consideration to the risk of contamination from other sources and other ingredients, giving an increased overall level.

(f) Polychlorinated biphenyls (PCBs)

PCBs are members of a group of organic compounds that have been used in a number of industrial applications. Because these compounds are toxic and environmentally stable, their use has been limited to closed systems, and their production has been banned in a number of countries. The most significant source of PCBs in food-stuffs is through absorption from the environment by fish. PCBs then accumulate through the food chain and can be found in high levels in tissues with high lipid content. This issue should be considered by HACCP Teams dealing with raw materials of marine origin.

(g) Dioxins and furans

Neither of these two groups is manufactured directly but they are created as by-products in the processes used to manufacture pesticides, preservatives and disinfectants, and in paper processing. They can also be formed when materials such as plastic, paper and wood are burned at low temperatures. There are several hundred dioxins and furans, some of which are non-toxic, some only slightly toxic and a small number are amongst the most toxic substances known. Dioxins are ubiquitous environmental contaminants and are generally present in very low concentrations in all foods.

A recall of poultry and poultry feed occurred in the USA in 1997 when dioxin was found in clay used as a 'free-flow' agent in the production of poultry feeds (*Food Chemical News*, 1997).

(h) Polycyclic aromatic hydrocarbons (PAHs)

PAHs are composed of benzene rings linked together and are the largest class of known environmental carcinogens. They are found in water, air, soil and food. They originate from coal-derived products, charcoal broiling, engine exhausts, petroleum distillates, smoke curing and tobacco smoke.

(i) Plasticizers and packaging migration

Certain plasticizers and other plastics additives are toxic and are of concern if they are able to migrate into food. Migration depends on the constituents present, and also on the type of food, for example fatty foods promote migration more than some other food-stuffs.

The constituents of food-contact plastics and packaging are normally strictly governed by legislation, along with the maximum permitted migration limits in a number of food models. The HACCP Team should be aware of current issues for both food packaging and plastic utensils, and should build control into the HACCP System. This might mean the requirement for checks on migration at the packaging concept stage.

(j) Veterinary residues

Hormones, growth regulators and antibiotics used in animal treatment can pass into food. Hormones and growth regulators have been banned from food production in many countries, and the use of antibiotics and other medicines are tightly controlled. Carry-over of antibiotics can cause major problems due to the potential for serious allergic responses in susceptible individuals. Similarly, hormones and growth regulators can cause toxic responses when consumed. The HACCP Team should consider the risks of contamination in their raw materials and product so that appropriate control and monitoring can be instigated. This will include control at the primary producer and may also involve monitoring at the incoming raw material stage.

(k) Chemical additives

Additives are used not only to make products safe and hygienic, but also to assist processing and to enhance or beautify what would otherwise be bland but nutritious products. They may also be beneficial, as in the case of vitamins.

The use of chemical additives is governed by regulation in almost all countries in the world. In Europe Directives 89/107/EEC (1989) and 95/2/EC (1995) classify additives according to their purpose (such as preservative, acidulant or emulsifier) and lay down guidelines and limitations for their use across various categories of foodstuffs. This, in effect, provides a positive listing of permitted additives. Therefore, if an additive appears in this or other countries' positive legislation, it may be assumed to have undergone appropriate toxicological testing and be deemed, by advisory committees of experts, to be safe. This testing procedure led to the European 'E' number system of classification for approved and tested materials and also to the RDA (Recommended Daily Allowance) levels which are set for such materials.

Nevertheless, it is still possible to imagine situations where careless or unnecessary use of additives poses a potential hazard in a foodstuff. For instance, in the choice of preservative one might avoid the excessive use of sodium metabisulphite in acidic products as the resultant sulphur dioxide gas may be injurious to asthmatics, both in the workplace and as consumers. Similarly, nitrates and nitrites could be avoided if suitable alternatives are available, without compromising food safety and quality. There are some synthetic colours, such as tartrazine, for which a causative relationship has been suggested, but not proven, for hyperactivity in children. It would obviously be sensible to use more natural alternatives in cases where the product is targeted at young consumers. Even then, it is wise to remember that 'natural' does not always mean 'safer'. Many natural plant extracts, for instance, are acutely toxic. Generally, materials can be used only if they are derived from normally consumed foodstuffs. Care must also be taken so that the 'natural additive' is not offered in amounts greatly in excess of those encountered in the native foodstuff. Additives may be beneficial, benign or, if misused, harmful. Great care and understanding must be exercised in their selection and use.

(l) Chlorophenols and chloroanisoles

Problems may be experienced in food and drink products with off-flavours described as medicinal and musty. In such cases the problems are usually due to contamination with chlorophenols or chloroanisoles. Although not a direct food safety issue, the consumer may be upset and may even feel unwell due to the unexpected, strange flavour or taste.

Chlorophenols are flavour active at concentrations in the low parts per million (p.p.m.) and impart disinfectant/TCP (trichlorophenol) like odours and flavours. These compounds are frequently constituents of cleaning agents and can also arise through chlorine bleaching of packaging materials where chlorination of the natural plant phenolics occurs.

Chloroanisoles are flavour active at concentrations in the low parts per billion (p.p.b.). They impart damp, musty off-flavours and are produced by microbial action on chlorophenols. Cardboard cartons are classic routes by which problems with chloroanisoles can occur. If the cartons become damp, microbial action on the very low levels of chlorophenols in the packaging can produce chloroanisoles which then diffuse from the packaging into the products within the cartons.

As for microbial hazards, chemical hazards continue to emerge as more information becomes available on the toxicity of various chemical compounds, and their likely occurrence in food. This is often covered by government research programmes, and may be followed by specific legislation.

Therefore, it is important that members of the HACCP Team, or their specialist advisors, remain fully up to date with the latest information on chemical compounds that may become the chemical hazards of the future.

4.2.3 Physical hazards

Physical hazards, like biological and chemical hazards, can enter a food product at any stage in its production. There is a huge variety of physical items that can enter food as foreign material, some of which may also be described as macrobiological, but only a few of these are hazards to food safety. Here we must ask ourselves very carefully whether or not any potential foreign material items are likely to cause a health risk to the consumer. Only if they are should they be considered in the main food safety HACCP Study. If, however, you choose to use the HACCP techniques separately in other areas, such as quality (Chapter 9), you may wish to extend your terms of reference to include all potential foreign material as hazards. It should be noted that you could be prosecuted in some countries (e.g. the UK) for the presence of foreign material in a product, regardless of whether or not it is a true safety hazard, but simply because the product is not of the true nature and substance demanded by the consumer. For this reason, recent trends have seen some manufacturers extending the scope of their HACCP studies to cover all potential foreign material. Alternatively, where the HACCP study is also extended to include additional issues, e.g. legal, then all these additional controls may be clearly separated from those that are critical to food safety (Chapter 9).

It is important to remember that any foreign material item could be a safety hazard if it has the potential to make the consumer choke. This is particularly important in foods that may be consumed by children, where even pieces of paper sacks or boxes could pose a safety risk. As with macrobiological hazards, it should also be noted that any foreign material item could transport microbiological hazards into the product, and this is particularly significant if they gain access after all processing steps that would control these hazards.

Your procedures for Good Manufacturing Practice and/or prerequisite programmes (Chapter 3) should ensure that these issues are considered as part of the building environment and should prevent any physical hazards from being brought into the production area by employees.

Foreign material items are food safety hazards if they fall into one or more of the following categories:

- items that are sharp and could cause injury;
- items that are hard and could cause dental damage;
- items capable of blocking the airways and causing choking.

Additional information to that listed here can be found in a number of publications, including CCFRA (1995b) and Haycock and Wallin (1998).

The main physical food safety hazards are as follows.

(a) Glass

Glass fragments can cause cuts to the customer's mouth and could have very serious consequences if swallowed. Smooth pieces of glass, e.g. watch glasses, could also cause injury by choking or could be broken into sharp pieces when the consumer bites into the product.

Glass may be present in the raw materials, e.g. as foreign material from the growing site, or may be the raw material container. Containers made from glass should be avoided wherever possible and should be kept out of the processing area. In addition, personnel should be prevented from bringing any glass items into production and sight glasses or glass gauges on equipment should be avoided. Glass light fittings should always be sheathed with plastic to prevent product contamination if the light shatters.

It may be that your finished product is filled into glass containers. In this case it is obviously not possible to keep glass out of the production area, but it must be properly managed and you should always have stringent breakage control procedures in place.

Another control mechanism for glass in food products is the use of X-ray detection devices, although these are currently not widely used due to expense and problems in application.

(b) Metal

Like glass, metal can enter the product from the raw materials or during production and can cause injury, as sharp pieces, or by choking. It is particularly important with this hazard issue that you ensure that your equipment is properly maintained so that parts do not drop into the product. All engineering work must be properly managed and parts, e.g. nuts and bolts, must not be left lying around. Where raw materials are delivered in metal containers, these should be opened carefully to minimize swarf contamination. This should be done outside the main production area if possible.

All products should be metal detected and/or passed over a magnet at least once, and this should be at, or as close to, the end of production and filling as possible. Where the finished product is held in metal containers, these should be adequately managed and product metal detection should take place immediately before product filling and closure. Metal detectors and magnets should be carefully chosen and calibrated to pick up the smallest pieces of each potential metal type.

(c) Stones

Stones are most likely to originate in raw materials of plant origin, where they may be present within the plant, e.g. between leaves, or be picked up during harvesting. They can cause the consumer dental damage or choking, and sharp stones may cause similar problems to broken glass and metal.

Stones can most easily be prevented by careful choice of raw material supply, and can be removed through the use of sieving/filtration, flotation tanks and centrifugal separators.

(d) Wood

Sharp splinters of wood could be a hazard to the consumer, causing, for example, cuts to the mouth and throat. Pieces of wood could also get stuck in the consumer's throat and cause choking.

Wood can enter the production area and the product in a number of ways. It may be present in raw materials, e.g. in plant material brought in from the fields, or it may be part of the raw material packaging. Wooden crates and pallets should be avoided where possible, and must not be allowed into production areas. Where wooden packaging or pallets have to be used for your own products, these must be carefully managed, and must not be allowed access to production areas where product is exposed.

Ideally all such wood should be contained in separate raw material handling and outer packaging areas. Production personnel must be prevented from bringing any wooden items into production areas. This should be part of every company's Good Manufacturing Practices or prerequisite programmes, and should be included in induction training for all staff.

Some products actually contain wood as one of their raw materials. These include ice-cream stick bars and traditional fish products such as herring rollmops. Obviously, here it is not possible to keep wood out of the production area, but it should be obtained from an approved source and handled in a controlled manner to prevent any splintering.

If you are operating from an old manufacturing site, it is possible that there is some wood built in to the processing area environment. Here you need to assess the risk of splinters breaking off into the product, but from a general hygiene point of view you should put together a plan for its removal and replacement. The HACCP Team will be able to use the HACCP techniques they are learning to help prioritize the essential areas for improvement.

(e) Plastic

Plastic is often used to replace other physical hazards, such as glass and wood, although it should be noted that hard plastic shards, e.g. from broken equipment guards, can also be hazardous. Soft plastic is also used as packaging or for protective clothing such as aprons and gloves. While more shatterproof than glass, you should implement similar breakage control procedures for hard, brittle plastic as for glass. For soft plastic, handling procedures and staff awareness are important, and soft plastic used during processing is often brightly coloured (usually blue) to assist with its identification.

(f) Pests

We have already considered pests as causes of biological hazards through the introduction of pathogenic micro-organisms into foods. Pests may also be thought of as physical hazards as their presence in foodstuffs may cause injury or choking. Most important here are large insects and parts of rodents or birds. An effective pest control programme must be in place to control these hazards on all food production, storage or preparation premises.

(g) Intrinsic material

Bones in meat/fish products, nut shells and extraneous vegetable matter would fall into this category. Control options include the use of X-ray detectors. However, this type of equipment can be costly and some industries therefore use careful sorting and inspection to minimize risk.

Many of these physical hazard issues can be controlled effectively as part of Good Manufacturing Practice procedures at your facility. If you already have these procedures in place properly, perhaps as part of site prerequisite programmes (Chapter 3) then the HACCP Team will be able to concentrate on the critical product contamination areas. Effectively this means that some hazards have been 'designed out' by the correct level of control in the prerequisite programme, and therefore need not be identified in the hazard analysis. This obviates the need to repeat the same foreign material hazards at every step in the process where environmental contamination is possible. Care should be taken here, however, to ensure that hazards are fully controlled by the prerequisite programmes, before they are dropped from consideration in the hazard analysis, and that the effectiveness of the prerequisite systems is fully verified.

4.3 Understanding control measures and control points

It is likely that you will have many controlling steps in your process, some of which are controlling the hazards we have just mentioned, and others which are not directly associated with control of safety. These will probably be points that are controlling the quality and legal attributes of your products, and are normally called manufacturing or process control and legal control points, respectively. Figure 4.1 illustrates the three types of control points.

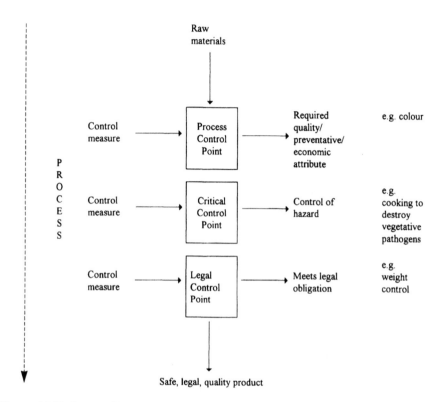

Figure 4.1 Understanding control points.

Critical Control Points (CCPs) are the stages in a processing operation where the food safety hazards must be controlled. These points ensure that the hazard cannot cause harm to the consumer of the finished product. In HACCP terms, a CCP is defined as follows:

CRITICAL CONTROL POINT:
A step where control can be applied and is essential to prevent, eliminate or reduce a food safety hazard to acceptable levels.

CCPs are essential for product safety, as they are the points where control is effected. However, the CCP itself does not implement control. Instead it is the action which is taken at the CCP that controls the hazard. These actions are normally described as control measures, although you may also see them written as preventative measures or control options. As we saw in Figure 4.1, process and legal control points will also require some sort of 'control measure' in place in order to effect control. However, as the scope of this book is HACCP and product safety, we will focus on the control measures for CCPs. In HACCP, control measures are the factors that eliminate or reduce the occurrence of hazards to an acceptable level.

CONTROL MEASURES:
Any factor or activity that can be used to prevent, eliminate or reduce to an acceptable level a food safety hazard.

Examples of effective control measures for different types of hazards are given in section 4.4.

Some companies implement extra hazard control points in their processes. These are normally upstream from the CCP for the particular hazard under consideration and are designed either to protect the process and process equipment, or to relieve pressure from the CCP by reducing the hazard. Examples are found in bulk operations where each unit of the process may have a hazard-controlling step such as metal detection or magnets. Although hazards are controlled at these points, they are not genuine CCPs as they are not the steps where control *must* be applied for product safety. These additional control points are often described as preventative or manufacturing control points, or simply Control Points (CPs), and their purpose is to give the process a greater degree of control. However, they should not be confused with the genuine CCPs where the hazards *must* be controlled.

It is important that CCPs are kept to the points which are truly critical to product safety, and this means that their number is usually kept to a minimum, specific to each unique process, in order to focus attention accurately on the essential controlling factors.

If you are having difficulties in telling the difference between CCPs and process control points, you should ask yourselves this simple question:

If I lose control is it likely that a health hazard will occur?

If the answer is yes, then the point must be managed as a CCP, and an effective control measure must be identified. (It may, of course, also be a legal control point.)

The effective operation of all CCPs is crucial to the safety of the product. In Chapter 6 we will investigate the tools and techniques for CCP identification.

Figure 4.2 'If I lose control, is it likely that a health hazard will occur?'

4.4 Practical hazard control

For any one hazard there may be a number of possible control measures. This will depend on the source or cause of the hazard or, for microbiological hazards, on the way the hazard is manifested. For example, *Listeria monocytogenes* may be a hazard in the raw materials for a cooked product. Here the hazard is **presence** of the organism and the control measure will be applied through the cooking step. The same micro-organism may also be a **cross-contamination** hazard to the same product after cooking, where the control measure(s) will be associated with prevention of contamination through environmental management and good handling practices. Similarly, the control measures for glass hazards in a raw material may be through effective supplier quality assurance, while the control for glass contamination during production will involve on-site glass management procedures, which may be part of a prerequisite programme.

Food products will be safe only when all the relevant hazards are controlled. In order to achieve this end, care must be taken in selection of appropriate control measures to operate at the CCPs. Examples of possible control measures for different types of hazards are listed in Table 4.2 in order to assist the HACCP Team.

Table 4.2 Examples of practical hazard control

Hazard	Control measures
Hazard category: all	• Prerequisite programmes/support systems, e.g. SQA, cleaning • Effective trace and recall procedures
Hazard category: biological Heat-stable pre-formed toxins, e.g. *Staphylococcus aureus, Bacillus cereus* emetic toxin	Raw materials • Specification for organism and/or toxin • Evidence of control during supplier process • Testing (positive release with statistically valid sampling) • Certificate of analysis (checked for compliance with specification) People • Handwash procedures • Covering cuts/wounds, etc. • Occupational health procedures • Management control of food handlers Build-up during process • Control of time that ingredients, intermediate and finished products are held within the organism's growth temperature range • Design of process equipment to minimize dead spaces • 'Clean as you go' procedures • Control of rework loops
Vegetative pathogens, e.g. *Salmonella* spp., *L. monocytogenes*, *V. parahaemolyticus, Y. enterocolytica*, *E. coli*	Raw materials • Lethal heat treatment during process • Specification for organism[a] • Evidence of control during supplier process[a] • Testing (as previous)[a] • Certificate of analysis (as previous)[a] • Temperature control to prevent growth to hazardous levels[b]

Table 4.2 *continued*

Hazard	Control measures

	• Intrinsic factors[b] such as pH and acidity; a_w – salt, sugar, drying; organic acids; chemical preservatives
	• Processes[b] such as irradiation, electrostatic field sterilization
	Cross-contamination at the facility (from the environment and raw materials)
	• Intact packaging
	• Pest control
	• Secure building (roof leaks, ground water, etc.)
	• Logical process flow, including where necessary:
	  (i) segregation of people, clothing, equipment, air, process areas
	  (ii) directions of drains and waste disposal

Spore formers, e.g. *Cl. botulinum*,
Cl. perfringens, *B. cereus*
	Raw materials
	• Specification
	• Evidence of control during supplier process
	• Certificate of analysis (as previous)
	• Lethal heat treating during process
	  (i) e.g. for *Cl. botulinum* $F_0 3$ process required for low acid products for ambient storage
	  (ii) Lethal combination of heat treatment and acidity or sugar level for high acid/sugar products for ambient storage
	  (iii) For products to be stored at chilled conditions (<5 °C) a sub-lethal heat treatment may be used but this must be accompanied by intrinsic factors which will prevent the growth of psychotrophic organisms (e.g. *Cl. botulinum*) during the product shelf-life
	  (iv) For all the above processes, pack integrity, cooling water chlorination and cooling container handling practices are critical
	• Temperature control to prevent growth to hazardous levels
	• Intrinsic factors such pH and acidity; a_w – salt, sugar, drying; organic acids; chemical preservatives

- Other processes lethal to the organism of concern, e.g. irradiation, etc.[c]
- Cross-contamination at the facility (from the environment and raw materials)
  - Intact packaging
  - Pest control
  - Secure building (roof leaks, ground water, etc.)
  - Logical process flow, including where necessary:
    - (i) segregation of people, clothing, equipment, air, process areas
    - (ii) direction of drains and waste disposal

Foodborne viruses, e.g. hepatitis A, SRSV

- Strict SQA control concerning irrigation and wash water of salads and vegetables, and sourcing of filter-feeding shellfish – avoidance of shellfish likely to be grown in sewage-contaminated waters
- Consideration given to proven lethal treatments such as irradiation or heat treatment
- Stringent personal hygiene procedures among food handlers

Parasites

- SQA procedures to include farm animal husbandry and veterinary inspection for control of parasites such as *Toxiplasma gondii*, *Taenia* in beef and pork, and *Trichenella* in pork
- Freezing (−18 °C), heating (>76 °C), drying and salting

Protozoa, e.g. *Cryptosporidium*, *Giardia*, *Cyclospora*

- Use of filtered water
- Pasteurization of raw milk
- Heat treatment of water and raw milk used as ingredients

Mould (mycotoxins), e.g. patulin, aflatoxin, vomitoxin

- SQA control of harvesting and storage to prevent mould growth and mycotoxin formation in cereals, groundnuts, dried fruit
- Heat treatment during process to destroy mould and prevent growth in product
- Controlled dry storage
- Intrinsic factors to reduce a_w to <0.7

Table 4.2 continued

Hazard	Control measures
Hazard category: physical	
Intrinsic physical contamination of raw materials[d]	Liquids
	• Filtering
e.g. Bone – meat/fish	• Magnets
Extraneous vegetable matter – fruit stones, stalks, pips, nutshells	• Centrifugal separation
	Powders
Extrinsic physical contamination of raw materials	• Sifting
	• Magnets
e.g. Glass	• Metal detection
Wood	• Air separation
Metal	Flowing particles, e.g. nuts, dried fruit, IQF fruit and vegetables
Plastic	• 100% inspection – electronic or human
Pests	• Screening
	• Sifting
	• Magnets
	• Metal detection
	• Washing
	• Stone and sand traps
	• Air separation
	• Flotation
	• Electronic colour sorting
	Large solid items, e.g. carcasses, fish, cabbages, cauliflowers, frozen pastry, packaging
	• X-ray detection
	• Metal detection
	• De-boners
	• Visual inspection
	• Electronic scanning

Physical process cross-contaminants,
e.g. Glass
- Elimination of all glass except lighting which must be covered – light breakage procedure
- Glass-packed products – glass breakage procedures, inversion/washing/ blowing of glass packaging before use

Wood
- Exclusion of all wooden materials such as pallets, brushes, pencils, tools from exposed product areas
- Segregation of all packaging materials
- Equipment design – preventative maintenance

Metal
- Avoidance of all loose metal items – jewellery, drawing pins, nuts and bolts, small tools
- Metal detection – sensitivity appropriate for the product, calibrated (3-monthly) and checked (hourly), ferrous, non-ferrous and stainless; fail-safe divert systems; locked reject cages; traceability

Plastic
- Avoidance of all loose plastic items – pen tops, buttons on overalls, jewellery
- Breakage procedures in place where hard brittle plastic is used

Pests
- Pest control programme:
  (i) Prevention, e.g. facility design, avoidance of harbourage areas, waste management, ultrasonic repellents
  (ii) Screening/proofing, e.g. strip curtains, drain covers, mesh on windows, air curtains, netting
  (iii) Extermination, e.g. electric fly killers, poisoning, bait boxes, traps, perimeter spraying, fogging

Building fabric
- Design and maintenance

Table 4.2 continued

Hazard	Control measures
Hazard category: chemical	
Cleaning chemicals	• Use of non-toxic, food-compatible cleaning compounds • Safe operating practices and written cleaning instructions • Separate storage for cleaning reagents • Covered designated labelled containers for all chemicals
Pesticides, veterinary residues and plasticizers in packaging	• Specification to include suppliers' compliance with maximum legal usage levels • Verification of supplier' records • Annual surveillance programme of selected raw materials
Toxic metals/PCBs	• Specifications and surveillance where appropriate As contaminants:
Nitrates, nitrites and nitrosamines and other chemical additives	• Specifications and surveillance where appropriate As additives: • Safe operating practices and written additive instructions • Special storage in covered, designated labelled containers • Validation of levels through usage rates, sampling and testing
Allergens/food intolerance	• Awareness of the potential allergenic properties of certain ingredients. Special consideration given to adequate labelling, production scheduling and cleaning, segregation or cross-contamination controls, rinse water testing, dedicated equipment, and to the control of rework

^a Essential when your process has no lethal heat treatment.
^b NB *Salmonella* spp. may cause infection at low numbers in your product. Therefore absolute confidence in your raw materials as supplied is necessary if there is no lethal process step. Remember also that heat-labile toxins will not necessarily be destroyed by other processes/controls such as irradiation or acidity.
^c Remember that heat-labile toxins will not necessarily be destroyed by other processes/controls such as irradiation or acidity.
^d NB Supplier Quality Assurance (SQA) procedures should include maximum acceptable levels in specifications. Sampling and visual inspection will supplement control measures.

5

Designing safety into products and processes

When designing a new food product it is important to ask if it is possible to manufacture it safely. Effective HACCP Systems will manage and control food safety issues on an ongoing basis but what they cannot do is make safe a fundamentally unsafe product.

It is essential that food safety is designed into a product at the development stage and this should be the responsibility of the Product Development and the HACCP teams working together. Possibly your HACCP Team will include a member of the Product Development Department who can introduce new product/process ideas at an early stage. There is no point in new product prototypes being shown to marketing departments or to customers if there are inherent safety risks which cannot be controlled. Not only will these be highlighted later, when the HACCP Study is carried out, leading to likely postponement of the product launch, but you may also be responsible for foodborne illness in the marketing or customer buying department.

Several factors must be considered when designing food safety into a product, and the HACCP Team and other relevant specialists must be involved at the outset. In this chapter we will consider the product formulation and process technologies, along with the importance of ensuring raw material safety. We will also discuss the establishment of a safe and achievable shelf-life, and finally, show an example of how this information may be organized into an individual product safety assessment.

5.1 Intrinsic factors

Intrinsic factors are the compositional elements of a food product and these can often have a controlling effect on the growth of micro-organisms. The major intrinsic factors found in foodstuffs and considered here are pH and acidity, organic acids, preservatives, water activity and the ingredients themselves. The information given is an introduction only and, where necessary, HACCP Teams should refer to specific and more detailed reference books.

5.1.1 pH and acidity

Acidity is often one of the principal preserving factors in food products, preventing the growth of many food-poisoning or food-spoilage organisms at certain levels. In fact, fermenting and acidifying foodstuffs are food preservation techniques that have been used for thousands of years. Examples of foods that can be preserved safely by pH and acidity are yoghurt, which is fermented to low pH by the action of starter cultures, and pickled vegetables, which are acidified with acetic acid (vinegar) and normally also pasteurized to prevent spoilage.

Although measurement of acidity is still often used in manufacturing for flavour control, the more useful parameter of measurement from the food safety viewpoint is that of pH. This is because published information on the growth and survival characteristics of micro-organisms at different levels of acidity is normally based around the pH scale.

There is a characteristic pH range across which micro-organisms can grow and the limiting pH for growth varies widely between different species. Most micro-organisms grow best at around neutral pH 7, but may also grow at values ranging from pH 4 to pH 8. A small number of bacteria can grow at pH < 4 or pH > 8 but those able to grow at pH < 4 are not normally associated with food poisoning. However, the growth of these acid-tolerant organisms could have food safety implications if their growth in the foodstuff is involved in raising the pH to a level where other micro-organisms, including pathogens, can grow. This is also true for yeasts and moulds which can grow at pH values considerably lower than pH 4.

It should also be remembered that micro-organisms may survive at pH values outside their range for growth. This has significance for food safety when other factors cause the pH to change. For example, spores of *Bacillus cereus* might be present in a low-pH raw material where they are unable to grow. If this is then mixed with other raw materials to make a higher-pH product, the spores may be able to germinate and grow to dangerous levels.

The pH limits for growth of a number of potential food pathogens can be found in Appendix B. The data shown in the Appendix are absolute limits for growth, many of which have been established in pure culture experimental studies. In real food situations the organisms may not be able to grow to these extremes for a number of reasons. These include water activity, oxygen concentration, heat or cold damage, and competing microflora. The effect of pH on growth is particularly affected by temperature and an organism that can grow at pH 4.5 at 30 °C may not be able to do so at 5 °C, and vice versa. The tolerance of micro-organisms to pH can also be greatly affected by the type of acid used.

5.1.2 Organic acids

Certain organic acids are widely used as preservative factors in food manufacture, although some of these are only permitted to be used in defined concentrations. The antimicrobial activity of organic acids is due to the undissociated molecules, although the exact mechanism of their action is unknown. The effectiveness of these acids is related to the pH of the sample, as the dissociation of the molecules is pH dependent. For example, the level of sorbic acid (usually added to a product as potassium sorbate) which will be effective in product at pH 7 will be only 0.48% compared with 97.4% in a product at pH 3. Tables 5.1 and 5.2 (adapted from ICMSF, 1980) illustrate the antimicrobial activity of organic acids and pH dependence.

Table 5.1 Percentage of organic acid undissociated at various pH values

Acid	pH value				
	3	4	5	6	7
Acetic	98.5	84.5	34.9	5.1	0.54
Citric	53.0	18.9	0.41	0.006	<0.001
Lactic	86.6	39.2	6.05	0.64	0.064
Benzoic	93.5	59.3	12.8	1.44	0.144
Sorbic	97.4	82.0	30.0	4.1	0.48
Propionic	98.5	87.6	41.7	6.67	0.71

Organic acids are most effective against micro-organisms in combination with other preserving factors, although there are several drawbacks to their use:

1. resistance of individual strains of micro-organisms to organic acids varies considerably;
2. organic acids are less effective if high levels of micro-organisms are present initially;

Table 5.2 Percentage of undissociated acid that inhibits growth of most strains

Acid	Entero-bacteriaceae	Bacillaceae	Micro-coccaceae	Yeasts	Moulds
Acetic	0.05	0.1	0.05	0.5	0.1
Citric	>0.005[a]	>0.005	0.001[b]	>0.005	>0.005
Lactic	>0.01	>0.03	>0.01	>0.01	>0.02
Benzoic	0.01	0.02	0.01	0.05	0.1
Sorbic	0.01	0.02[c]	0.02	0.02	0.04
Propionic	0.05	0.1	0.1	0.2	0.05

[a] Actual inhibitory concentrations likely to be far in excess of these values.
[b] This value is for *Staphylococcus aureus*; micrococci are more resistant.
[c] Clostridia are more resistant.

3. micro-organisms may become resistant to their use;
4. they can be utilized as carbon sources by many micro-organisms.

Organic acids commonly used as preserving factors in foods include acetic, citric, lactic, benzoic, sorbic and propionic acids. Acetic, citric and lactic acids are often added as part of the formulation from the flavour point of view, while benzoic, sorbic and propionic tend to be used only for their preservative action. The specific organic acid chosen depends on the target microflora for inhibition, along with the formulation and other intrinsic factors present in the foodstuff.

5.1.3 Preservatives

Chemical preservatives can be added to certain foodstuffs to inhibit the growth of food poisoning and spoilage organisms. There are usually carefully controlled legal limits for addition, and different preservatives are effective against different groups of micro-organisms. Examples of preservatives commonly used in foods are sodium nitrite, which is often used in cured meat products, and potassium sorbate, which is used in many areas, including bread, cake and jam manufacture. Other food preservatives include nisin, sodium nitrate, sulphur dioxide, sodium benzoate, sodium and calcium propionate, and sodium metabisulphite. Some of the commonly used chemical preservatives are the soluble salts of the previously mentioned organic acids.

Sodium nitrate and nitrite have long been used in meat curing to reduce spoilage and stabilize colour. Their safety effect is to prevent the germination of spores, thus controlling pathogens such as *Clostridium botulinum*. Their effectiveness depends on a number of factors, including the types and numbers of micro-organisms present, the curing temperature and the meat pH.

As we saw in the previous section, potassium sorbate or sorbic acid is effective in acid foods, particularly against yeasts and moulds. It will also limit the growth of micrococci, entero-bacteriaciae and bacilli, although not clostridia. Similarly, sodium benzoate or benzoic acid is also effective, mainly in high-acid foods. It will inhibit the growth of yeasts and moulds and is commonly used in pickles, salad dressings and fruit juices.

Nisin is an antibiotic which prevents the growth of many bacteria, and which has been used in cheese manufacture and canned foods. This preservative tends to be relatively expensive and this has limited its application. Sulphur dioxide is an antioxidant which inhibits the growth of bacteria and moulds, and which can be used in gaseous or liquid form. It is commonly added to beers and wines, and to comminuted meat products. The propionates, sodium and calcium, are used to control moulds in low-acid foodstuffs such as cakes and bread.

The smoking of food also has a preservative effect due to chemical compounds present in the smoke. Although the exact mechanism of preservative action is poorly understood, smoking is a traditional method of food preservation which has remained popular, e.g. for smoked salmon. In recent years it has become fairly common to add smoke flavour to food rather than using the smoking technique. This has little or no preservative effect.

5.1.4 Water activity

Water activity (a_W) is a measure of the availability of water in a sample. As micro-organisms can only grow in the presence of an available form of water, they can be controlled by controlling the a_W. The a_W is the ratio of the water vapour pressure of the sample to that of pure water at the same temperature:

$$a_W = \frac{\text{water vapour pressure of sample}}{\text{pure water vapour pressure}}$$

Pure water has an a_W of 1.0, and as solutes are added making a more concentrated solution, the vapour pressure decreases and along with it the a_W. The a_W is directly related to the equilibrium relative humidity (a_W = ERH/100), as well as to the boiling point, freezing point and osmotic pressure of the sample.

Traditionally, a_W has been used as a preservative factor against micro-organisms in foods through the addition of salt and/or sugar and the reduction of moisture content through drying. Sugar has traditionally been added to fruit products, such as jams and soft drinks, while salt has wide application in products such as pickled, salted fish and dry cured meats. The minimum a_w values permitting growth for a number of food pathogens is given in Appendix B.

5.1.5 Ingredients

The individual ingredients and their interactions with each other should also be considered as intrinsic factors. Particular attention needs to be paid to hazards entering the product in this way:

- Do the ingredients contain hazards? For example, *Salmonella* in raw milk.
- Would the wrong quantity of an ingredient be hazardous? For example, too much or too little of a preservative or acid added.
- Are any ingredients allergenic? For example, nuts.
- Could ingredient interactions cause a hazard? For example, by neutralizing the preserving acid.

Recent trends have been to design products which have fewer inherent preservation factors, for example less sugar, less salt, less fat and no preservatives. As this affects the stability and safety of the product, the HACCP Team should be aware that the significance of safe raw materials and control during processing has increased.

5.2 Raw materials

In order to make safe products you must understand the hazards and risks associated with your raw materials. The raw materials should either contain no hazards, or any hazards present must be controllable by the process. This can be achieved through a planned and managed programme of Supplier Quality Assurance (SQA) (Chapter 3).

In establishing the level of control required for each of your raw materials, it is important to think about how they will be handled and processed. The same raw material may require different levels of control for two different products, e.g. herbs going into a cooked product may require less emphasis on microbiological control at the raw-material stage than the same herbs being used as a garnish on a ready-to-eat product. In order to assist in identifying the level of control needed a question–decision tree has been developed (Figure 5.1).

Following the sequence of questions in the decision tree will give you an understanding of the level of control required for each of your raw materials. Work through the decision tree as follows.

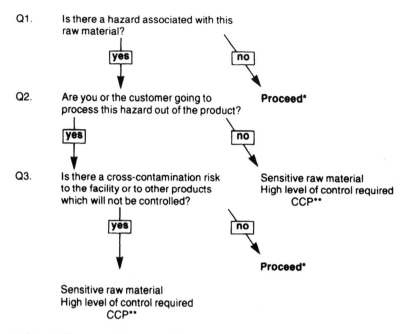

Q1.   Is there a hazard associated with this
      raw material?

      yes   no

Q2.   Are you or the customer going to   **Proceed***
      process this hazard out of the product?

      yes   no

Q3.   Is there a cross-contamination risk   Sensitive raw material
      to the facility or to other products   High level of control required
      which will not be controlled?   CCP**

      yes   no

                                       Proceed*

      Sensitive raw material
      High level of control required
              CCP**

* Proceed to your next raw material

** Following the hazard analysis, you are likely to find that this raw material must be
   managed as a CCP

Figure 5.1 Raw Material Control Decision Tree.

Q1 Is there a hazard associated with this raw material?

This first question may be obvious but it focuses the mind on identification of all food safety hazards associated with the raw material. If no hazards are identified, then you should move on to the next raw material, but if hazards are identified you should consider Question 2 for each one.

Q2 Are you or the consumer going to process this hazard out of the product?

If the answer to this question is no, then you could potentially have the hazard in your finished product, if you do not implement effective control at the raw material stage. This is therefore a very sensitive raw material and must be subjected to a high level of control, probably as a CCP in your process. If, however, the answer to this question is yes, then you should go on to consider Question 3.

Q3 Is there a cross-contamination risk to the facility or to other products which will not be controlled?

This question investigates whether the hazard could be carried through to your products by direct cross-contamination or via cont-amination of the facility. This is particularly important in a facility where several different products are being made, as there may be a lethal step built in to control the hazard in its intended product but not in other products it might cross-contaminate. If the answer to this question is no, and there is no cross-contamination risk, or any risk will be controlled effectively by existing mechanisms (e.g. plant layout), then you should move on to the next hazard or raw mater-ial. If the answer is yes, then control of the sensitive raw material must be effected here, and it is likely that this will be a CCP in the HACCP System.

Using the Raw Material Decison Tree will allow you to target SQA resource at the raw materials that are most critical to your operation and products. These raw materials should then be managed through your prerequisite SQA system.

5.3 Process technologies

There is a wide variety of different process technologies available and it is necessary that the type of process being used is fully under-stood.

It is essential that any planned **thermal processes**, in terms of heating, cooling and holding temperatures and times, are known, along with their effect on potential hazards. Where the consumer is expected to cook or heat a product, the exact instructions to achieve the desired heat process should be determined. Often product development is carried out on samples manufactured in a labora-tory or pilot kitchen. Where this is true, the process requirements to achieve the correct heat profile when scaled up to the manufactur-ing environment must be understood. This will vary depending on the type of heat processor chosen, e.g. band oven, plate heat exchanger, rack oven, bulk vessel, microwave, etc.

Where a product is being made by **fermentation**, it is important to understand the chosen culture system and how it is controlled. Would you know if the fermentation failed, and would this allow microbiological hazards to grow?

In the production of a **dried product**, how is the final moisture controlled? The potential for the presence of microbiological hazards that have survived through the process or have entered through contamination must be established, as these could cause a problem when the product is reconstituted. This is particularly important for products that are reconstituted without further heating, e.g. dessert mixes.

In a **freezing process**, the length of time to freeze and any holding stages before freezing could be significant. The potential cross-contamination risk is important here also, particularly for foods to be consumed immediately after defrosting. If you are using frozen ingredients in a product you will need to consider whether or not these need to be defrosted before addition. Adding frozen ingredients may help with temperature control, but they may change the product's intrinsic factors as they defrost, e.g. diluting the dressing of a low-pH salad and raising the pH.

Irradiation may be used to vastly improve the microbiological quality of certain foodstuffs. However, if the product has been mishandled before the irradiation process, it is possible that microbial toxins could be present which would not be affected by the process. This could cause a major food-poisoning risk in the finished product.

Will it be a **continuous process** or will there be a number of holding or delay stages? The maximum holding time or delay at all stages should be understood, along with the associated temperature at this stage. This information will need to be assessed during the hazard analysis to establish the potential for growth of microbiological hazards.

The chosen **packaging system** may have an impact on product safety, so the influence that packaging has on the growth of microorganisms during the product shelf-life should be established. The use of controlled and modified atmosphere packaging systems has increased in recent years, along with that of vacuum packaging. The absence of oxygen means that only anaerobic or facultative organisms can grow, and so these systems have been promoted for the extension of product shelf-life by reducing/preventing the growth of the normal microflora. However, they allow the growth of a different microflora which could include food pathogens. It is vital that these organisms cannot grow to hazardous levels during the proposed shelf-life.

If any new or less well-established process technologies are employed, then the hazards associated with these must be determined for full consideration during the hazard analysis. An example here is ohmic heating, which allows the sterilization of liquid-based foods without overcooking the liquid phase. Here a voltage is applied between electrodes inserted in a tube though which a continuous stream of food passes. The food is sterilized by the heat generated in it due to its electrical resistance. Initial considerations of this technique suggested that toxicological hazards might be formed by metal ions from the electrodes leaching into the food, or by the formation of free radicals in the food through the heating process. Detailed examination of the technique by expert toxicologists found that free radicals were not likely to be formed, and that any traces of metal in the food from the electrodes would not represent a hazard to health. The technique was therefore cleared for these hazards.

5.4 Establishing a safe and achievable shelf-life

When you are designing your products, you will need to consider the shelf-life that you and your customers would like for each product, and then go on to establish whether or not this proposed shelf-life is safe and achievable. Criteria that can influence your product's shelf-life include:

- raw materials;
- process technology used;
- product intrinsic factors;
- type of packaging;
- conditions during storage, distribution and retail;
- customer storage and handling.

The shelf-life will be limited by factors that cause the product to become unsafe or deteriorate, and these will be influenced by the criteria listed above. As we are concentrating on safety in this text, we will consider here only factors that cause the product to become unsafe. However, the HACCP techniques can be used to predict product deterioration or spoilage, as we will see in Chapter 9. Further information on determining shelf-life can be found in guidelines published by the UK Institute of Food Science and Technology (1993).

If you are a small manufacturer of high-risk products, you may wish to consider use of external experts to help with shelf-life determination.

5.4.1 What factors could cause the product to become unsafe?

The main factors that can cause products to become unsafe during their shelf-life are pathogenic micro-organisms. Rancidity of fats can cause revulsion and sickness when consumed, but these are normally associated with spoilage rather than safety.

We have already discussed microbiological pathogens as hazards in Chapter 4, and have looked at intrinsic factors earlier in this chapter. The pathogen profiles in Appendix B may be helpful in deciding whether the product is likely to provide an environment favourable to pathogen growth. Pathogenic micro-organisms may be present in your product from the raw materials, or from contamination during processing. These may be able to grow, depending on the intrinsic factors and packaging, along with the storage, distribution and handling conditions to which the product is subjected.

If you consider that a pathogen is likely to be present in your product, and that it will not be prevented from growth by the product's intrinsic factors, then you will need to investigate the degree of growth that is possible in the product. This, along with knowledge of the infectious dose for the organism in question, can be used to evaluate when the product will become potentially unsafe for consumption, and thus to limit the shelf-life to a safe level. It is important to note that if pathogens with a low infectious dose are likely to be present at the start of shelf-life, and the product is not due to be cooked thoroughly by the consumer, then the product is potentially unsafe and should be redesigned.

5.4.2 How do you know when pathogens reach unsafe levels?

Information on growth potential in foods, and with varying proportions of inhibitory intrinsic factors, can be found in the scientific literature. This can give you a good idea of the likely position in your product but should not be relied on absolutely for a safe shelf-life. Mathematical modelling of pathogen growth in various concentrations and combinations of intrinsic factors can also be carried out. A number of computer models have been developed which can be used or accessed by the HACCP Team and expert consultants, the major one available in the UK being the Food MicroModel.

The theoretical safe shelf-life obtained from literature values or mathematical modelling should be confirmed in practice for the product in question. This can be done through examination of the product for each micro-organism of concern throughout and beyond the proposed shelf-life. Product samples should be held under the expected storage and handling conditions, and it is prudent to build in an element of abuse, e.g. elevated temperature storage, to reflect possible product mishandling.

Where the micro-organism(s) of concern may not be present all the time, or may be present at very low levels, it is more appropriate to carry out product challenge testing to evaluate potential for growth. Here each individual pathogen is inoculated into the product, which is then held at the expected storage and handling conditions. As for standard shelf-life examination, the product is tested at various intervals throughout and beyond the proposed shelf-life and an element of abuse should be built in.

It is important to note that shelf-life should always be confirmed on product samples which have undergone the same treatment as all product which goes on sale. This means that any shelf-life proposed through theoretical studies, or through examination and challenge of development samples, must be verified on product which has been manufactured on the main production line, under the normal manufacturing conditions.

5.5 Product safety assessment

Companies which choose to use the modular approach for their HACCP System will need to ensure that individual product safety has been properly assessed. In this type of HACCP System, the HACCP Plans cover **processes**, which normally means that a number of different products are included within each module. It is important not to forget to consider each individual product through a form of product safety assessment. This is intended to pick up any product-specific hazards and can be used to document recommendations to the HACCP Team. Product safety assessments may be carried out either before HACCP Plan development, where all existing product varieties will be assessed, or after HACCP implementation, when new varieties are added to a product range. In the latter case it is particularly important to ensure that the existing HACCP Plan is still valid for the new product.

Throughout the remainder of the book we will be using a fictitious example of ice-cream manufacture to illustrate the design and implementation of a HACCP System. This example assumes that the manufacturer is a medium-sized company producing a number of different varieties and operating to acceptable food industry standards. The products are packed in family-sized and individual tubs for retail sale. Here, at the development stage, we introduce the chocolate-chip ice-cream product which will be managed by the modular HACCP Plan developed in Chapter 6.

5.5.1 Product concept safety considerations

Product safety assessment may be carried out and documented in a number of ways, but will usually comprise consideration of:

• target audience and food sector;
• raw materials;
• recipe and intrinsic factors;
• process conditions and cross-contamination issues;
• distribution and final customer/consumer handling.

The Product Development personnel can be helped by the HACCP Team in establishing the product criteria early on in the design of a new product. For existing products, the HACCP Team may wish to use this structured approach to consider likely hazards with each particular product type.

5.5.2 Development specification

In many cases the product safety assessment will be based on a development specification, such as the example given in Figure 5.2 for chocolate-chip ice-cream.

In most cases the suppliers of the raw materials will be known and outline or full specifications may be available. This is important as the next step in the assessment is the evaluation of likely raw material hazards.

Iced Delights Ice-Cream Manufacturers
Development Specification:
Chocolate-Chip Ice-Cream

Recipe

Skimmed Milk Powder (SMP)
Cream (40% fat)
Liquid sugar (80° Brix)
Milk chocolate chips (5 mm discs)
Water
Vanilla flavouring (liquid)
Stabilizer (lecithin)

Packaging

1 litre plastic tubs
500 ml waxed cartons
100 ml waxed cartons
Sealing film
Lids
Plastic spoons

Target Consumers

A high quality ice-cream for family or individual use.

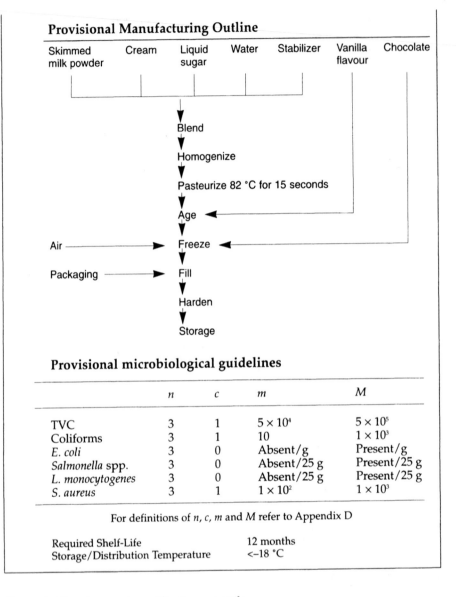

Provisional Manufacturing Outline

| Skimmed milk powder | Cream | Liquid sugar | Water | Stabilizer | Vanilla flavour | Chocolate |

Blend

Homogenize

Pasteurize 82 °C for 15 seconds

Age

Air ⟶ Freeze

Packaging ⟶ Fill

Harden

Storage

Provisional microbiological guidelines

	n	c	m	M
TVC	3	1	5×10^4	5×10^5
Coliforms	3	1	10	1×10^3
E. coli	3	0	Absent/g	Present/g
Salmonella spp.	3	0	Absent/25 g	Present/25 g
L. monocytogenes	3	0	Absent/25 g	Present/25 g
S. aureus	3	1	1×10^2	1×10^3

For definitions of n, c, m and M refer to Appendix D

Required Shelf-Life	12 months
Storage/Distribution Temperature	<–18 °C

Figure 5.2 Development specification example.

5.5.3 Raw material evaluation

Considerations of ingredient sensitivity during the design of a new product can assist in targeting the SQA activities to work with new suppliers at an early stage. A number of issues are likely to be discussed, for example:

Sensitivity status:

Why is the ingredient sensitive?

What foreign body hazards exist?

What, precisely, are the microbiological hazards?

Have the following chemical hazards been considered: pesticides, mycotoxins, processing aids, allergens, compound ingredient components?

Supplier control:

What is the approval status of the factory or agent?

What is the specification status?

Are certificates of analysis required?

Have previous audit reports been considered?

Are there any shelf-life criteria associated with ingredients?

Bearing this in mind, let's now look at the ingredients and packaging for this ice-cream, using the raw material decision tree as previously described (Table 5.3). The table shows how the information may be organized as each ingredient is evaluated.

Table 5.3 Chocolate-chip ice-cream – raw material decision matrix

Raw Material	Q1	Q2	Q3	CCP?	HACCP Team notes
Skimmed Milk Powder (SMP) — Salmonella	Y	Y	N	No	When we consider SMP, the answer to Q1 is Yes because of associated risks of salmonella. However, the answer to Q2 is also Yes as this ingredient will undergo a heat process which is lethal to vegetative pathogens. There is no cross-contamination risk at this facility as there is already full segregation of the raw materials before pasteurization from the post-process area, and from other sensitive raw materials such as chocolate chips. This raw material therefore does not require to be managed as a CCP at the SQA stage for this hazard
— Foreign material	N	—	—	No	Foreign material is not normally associated with SMP because the milk is filtered before drying and powder is sieved immediately before bagging
— Antibiotic residues	Y	N	—	Yes	Antibiotic residues may carry through to the final product and will not be removed by the heat process. So, as part of SQA, the raw milk supply into the dairy must be monitored

Hazard				Reasons
Cream – Vegetative pathogens (e.g. *Salmonella*, *Listeria*, *E. coli*)	Y	Y	N	No	This hazard is most likely to occur through post-process contamination, e.g. through poor tanker hygiene. However, the answer to Q2 is Yes and Q3 No, for the same reasons as SMP
– Foreign material	N	—	—	No	The answer to Q1 is No as there is an in-line filter in place at the supplying dairy
– Antibiotic residues	Y	N	—	Yes	As per SMP
Liquid sugar	N	—	—	No	No hazards were identified
Milk chocolate chips – *Salmonella*	Y	N	—	Yes	For chocolate chips there is a hazard of *Salmonella* being present. The chocolate chips will be added to the ice-cream after the heat process, and the consumer will eat the product without any further preparation. This leads us to the decision that a high level of control is required with this raw material, i.e. it will be a product CCP, and we should focus SQA resource here accordingly

Table 5.3 continued

Raw Material	Q1	Q2	Q3	CCP?	HACCP Team notes
Milk choc chips (cont.) – Chemical – pesticide residues	N	–	–	No	These hazards could occur at the growing and raw material storage stages. However, this would be routinely controlled through the prerequisite SQA programme
Water – Protozoa	Y	N	–	Yes	As an ingredient in this product there would be minimal risk from bacteria due to the heat process. However, the temperatures may not be sufficient for protozoan parasites such as Cryptosporidium
– Chemical, e.g. toxic metals, pesticides, nitrates	N	–	–	Yes	As an ingredient, control of the supply is critical as these hazards may not be processed out. However, this would be routinely controlled through the prerequisite SQA programme
Vanilla flavour – Microbiological	N	–	–	No	The processing by the supplier will eliminate any risk of either microbiological or physical hazards
– Physical	N	–	–	No	
Stabilizer (lecithin)	N	–	–	No	No hazard identified.

| Plastic tubs and film
— Chemical (plasticizers and additives) | N | N | Yes | There may be chemical leaching into product. The SQA process must ensure that all chemical constituents are legal and within chemical migration limits for a high-fat ice-cream product |
| Waxed cartons, waxed lids, plastic lids | N | — | No | No hazard identified |

5.5.4 Additional considerations

Other issues with potential product safety implications may include:

1. Legal constraints – these may not strictly relate to product safety, but it is useful to be aware of relevant legislation:
   - ingredient usage concentrations, e.g. maximum usage restrictions;
   - product compositional requirements;
   - processing requirements, e.g. pasteurization.

2. Recipe/intrinsic factors
   - Which intrinsic factors control food safety and at what levels?

3. Process conditions
   - Does the process affect the safety intrinsic factors?
   - Does the process make the product safe and why?
   - Are any hazards likely to be introduced due to the process?

4. Cross-contamination
   - Are there any obvious risk factors from or to existing products, packaging and the process environment?
   - Allergen control in addition to pathogen contamination would be an appropriate consideration.

5. Intended shelf-life
   - How susceptible is the product to abuse?
   - What governs the shelf-life?

6. Distribution
   - Is the product susceptible to damage or abuse?

7. Customer/consumer
   - Could additional hazards be introduced?
   - Is control necessary for any hazards at this stage?
   - Although not normally considered as a food safety hazard, could packaging cause health and safety hazards, e.g. injury while opening cans?
   - Do you understand all the potential uses of the product, e.g. in different recipes, etc.?

This information could be recorded in report format or using a simple table, as in the following example (Table 5.4).

When you have established the safety of your product design, and decided on the likely shelf-life, you can move on to look at how safety will be controlled from day to day during manufacture. This is through the establishment, implementation and maintenance of a HACCP Plan for the process, which we will begin to consider in Chapter 6.

Table 5.4 Product safety assessment – chocolate-chip ice-cream

Example

PRODUCT Chocolate-chip ice-cream			FORMULA			DATE 5-11-96		Page 1 of 4
Stage	Considerations	Criteria	Likely hazard	Control measures	Is control possible?	Validation of control	Recommendations to HACCP Team
Concept	Targeted at general population including high-risk groups. Domestic use	Frozen product to be eaten without any further process	Vegetative pathogens with low infective dose	Pasteurization, filtration, supplier assurance	Yes		
Ingredients	Sensitive ingredients and supplier control						
	SMP	Dried	Antibiotic residues	SQA	Yes	Supplier's antibiotic monitoring procedures	Verify that antibiotic monitoring procedures satisfactorily covered during SQA audits
	Cream	Pasteurized, chilled		SQA	Yes		
	Chocolate chips	Ready to use	Salmonella Pesticide residues	SQA	Yes	Certificates of Analysis received with each batch	Careful control at supplier → effective supplier management (including audit of processing site and microbiological test facilities)

Table 5.4 *continued*

Stage	Considerations	Criteria	Likely hazard	Control measures	Is control possible?	Validation of control	Recommendations to HACCP Team
Ingredients (cont.)	Water	Mains	Protozoa	Supplier control	Unknown	Legal obligation	Ensure proactive relationship with water authority
	Stabilizer	White Powder	No hazard identified	Labelling in plant	–	–	Controlled labelling of all ingredients must be in place in the factory
	Packaging	Plastic tubs and film	Plasticizers and additives	SQA	Yes	Supplier testing results	Ensure product suitability testing has occurred and is documented as complying with legal requirements
Legal	Ingredients/ product	Thermal process control recipe	Food safety		Yes	Regulations as per manufacturing country	Check compliance
Recipe/ Intrinsic factors	a_w, pH, chemical preservatives, organic acids – none will control product safety	Insufficient sugar to prevent microbial growth totally	No – product is frozen	–	–	–	–

PRODUCT Chocolate-chip ice-cream   FORMULA   DATE 5-11-96

Process	Process conditions					
	Pasteurization failure	Survival of vegetative pathogens	Correct heat process	Yes	Required	Ensure that the effectiveness of the heat process is validated for this formulation. Critical limits will need to be established
	Temperature control during ageing	Spore outgrowth	Effective temperature control and stock rotation	Yes	Audited on a monthly schedule. Calibrated temperature recording already in place	None
	Contamination	Air filtration failure	Introduction of pathogens	Effective filtration	Yes	Required	Check filter size and performance criteria. Microbiologically filtered air necessary
Post-factory	Shelf-life	Product consumed beyond shelf-life	No hazard identified	—	—	—	—

Table 5.4 *continued*

	PRODUCT Chocolate-chip ice-cream		FORMULA				DATE 5-11-96	
Stage	Considerations	Criteria	Likely hazard	Control measures	Is control Possible?	Validation of control	Recommendations to HACCP Team	
Post factory (cont.)	Customer abuse	Temperature abuse	Unlikely – sufficient abuse for growth will render product inedible	–	–	–	–	
		Contamination with serving spoon	Unlikely – only low numbers; will not grow in freezer	–	–	–	–	
			Slight risk perhaps from leaving serving spoons in water between servings	None possible	No	–	The product is targeted to the domestic market rather than catering, therefore hazards associated with mass servings are unlikely to be realized. Revisit if a 'catering' version is launched	

Signed: *J. Smith*   (Position) Development Manager   Date: *21-11-96*

6

How to do a HACCP Study

Now that we have discussed safe product design, we are ready to look at how to carry out the HACCP Study. In this chapter we will be identifying the potential hazards associated with products during processing and exploring the options for their control. In doing this, we will be looking at a number of useful techniques, such as brainstorming and the use of hazard analysis charts, which will help the HACCP Team to structure their approach. We will then move on to the identification of CCPs and start to build up the information required in the HACCP Plan – critical limits, monitoring procedures, corrective action and responsibility. This covers the requirements of HACCP Principles 1–5. We will continue to use the fictitious example product, chocolate-chip ice-cream, and will look at the construction of a modular HACCP Plan for an ice-cream manufacturing business.

6.1 What is the HACCP Plan?

The HACCP Plan is a formal document that pulls together the key information from the HACCP Study, and holds details of all that is critical to food safety management. The HACCP Plan is drawn up by the HACCP Team and consists of two essential components – the Process Flow Diagram and the HACCP Control Chart – along with any other necessary support documentation. It is important that the HACCP Plan is focused on food safety management and therefore additional documentation should be kept to a minimum. However, it is often useful to include a product and process description, and, if a Product Safety Assessment has been documented, then these may be filed together. Details of record keeping and verification procedures may also be included, although these could be held within the Quality Management System. You may also find it helpful to retain all preparatory documentation used by the HACCP Team which illustrate the hazard analysis thought process, although this documentation need not be part of the formal HACCP Plan.

6.1.1 The Process Flow Diagram

The application of HACCP Principle 1 begins with the construction of a comprehensive Process Flow Diagram. This is a stepwise sequence of events through the whole process, giving a clear and simple description of how the end product is made. It is an essential part of the HACCP Plan which enables the HACCP Team to understand the production process, and is the basis for the hazard analysis. It includes details of all ingredient handling procedures and follows the process through to the consumer. Consumer actions may also be included, depending on the terms of reference drawn up by the HACCP Team (section 6.2).

At the end of the HACCP Study all Critical Control Points (CCPs) identified are normally highlighted on the Process Flow Diagram, thus tying it together with the HACCP Control Chart. The Process Flow Diagram is also useful in demonstrating control of food safety to customers and regulatory inspectors.

6.1.2 The HACCP Control Chart

The HACCP Control Chart contains details of all the steps or stages in the process where there are CCPs. It is normally documented as a matrix or table of control parameters, and contains details of the hazards and control measures associated with each CCP, along with the control criteria and responsibilities.

In order to put together a HACCP Plan we use the HACCP Principles and follow a number of steps as detailed in Figure 6.1.

6.2 Define your terms of reference

When your HACCP Team is ready to start its first HACCP Study, it is important to agree on the terms of reference or scope before they begin. It is essential that the correct focus is established to prevent the team being bogged down in unnecessary detail.

HACCP was originally designed as a food safety management tool, and food safety should be your initial focus. However, as this is a very wide area in itself, the HACCP Team must decide first where to start, and also, just as important, where the study will end.

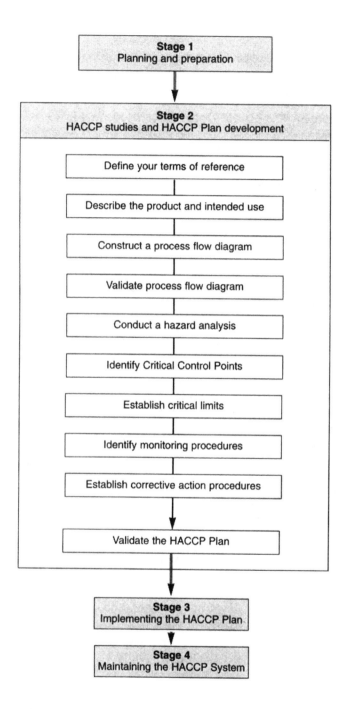

Figure 6.1 HACCP Key stage 2 — HACCP studies and HACCP Plan development.

There are a number of questions to help with these decisions:

Do you want to cover all types of hazards initially (i.e. microbiological, chemical and physical) or just one type, e.g. glass or microbiological?

Experienced HACCP Teams usually look at all types of hazards at once, and this is certainly better from the time management point of view. However, an inexperienced HACCP Team may find it easier to limit the number of hazard types in the initial study. The process can be revisited again afterwards to look at the other hazard types.

This approach may also be necessary if specialists, e.g. the microbiologist, are only available at certain times.

Will the study cover a whole process or one specific part, and is this for one or a group of products? That is, what are the start and end points?

This will depend largely on how you have planned your HACCP System. As we discussed in Chapter 3, you will have considered the length and complexity of the process, and whether it divides logically into separate process modules. However, it is worth re-emphasizing that, when process modules are fitted together, all process steps must be covered to ensure that no hazards are missed. It is particularly important to investigate what happens to the product when it moves from one process area to the next.

If the process being studied is common for a number of products, then these will be included in the scope, but it is essential that no hazards arising from slight differences in product formulation are overlooked. Potential issues should be highlighted in the individual Product Safety Assessment (Chapter 5).

You will want to consider whether primary producers will be included, e.g. if the majority of your raw materials are sourced directly from the agricultural producer, and also, whether the HACCP Study should continue through distribution, retail and consumer handling. To answer the former question you will need to refer to the outcome of the raw material decision tree and determine whether the suppliers have the capability to draw up and implement their own HACCP Plans.

To answer the latter question you will need to consider whether your product is safe at the end of production, i.e. all hazards have been controlled, or whether the product needs special handling. Is it a perishable product which could potentially be rendered unsafe by improper handling, or are you actually relying on consumer action to control any hazards, e.g. in a raw meat product?

When you have answered the above questions you will be able to define your terms of reference for the HACCP Study (Box 6.1).

Box 6.1 Terms of reference: ice-cream manufacture

This HACCP Study considers biological, chemical and physical hazards throughout the entire process.

Biological hazards include vegetative pathogens such as *Salmonella* and *Listeria* and toxin-formers such as *Staphylococcus aureus*. Chemical hazards could be associated with the raw materials, e.g. pesticides, antibiotics and allergens, or with contamination during the process, e.g. allergens cross-contamination.

The HACCP Team considered that a wide range of physical hazards would affect the safety of these products, as they are likely to be consumed by small children who may be susceptible to choking on large items.

As the ice-cream is to be sold as retail tubs and its safety is unlikely to be affected by storage and distribution, the HACCP Study stops at the dispatch stage. The HACCP Team had decided on a modular approach covering all products produced at the factory.

At this stage in the HACCP Process, the Iced Delights HACCP Team also documented the structure of their modular HACCP System (Figure 6.2). The shaded boxes in this figure are the modules that will be covered in our worked example.

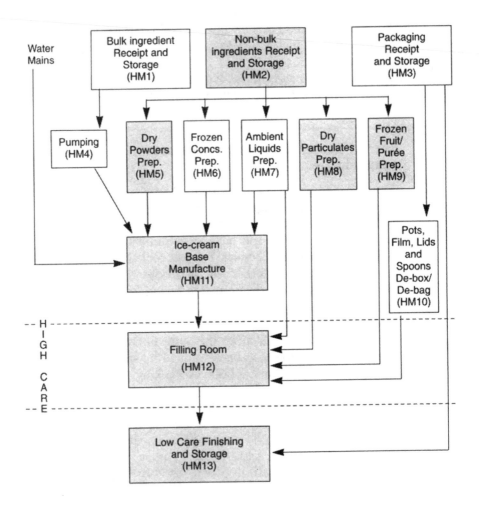

Figure 6.2 Ice-cream modular HACCP System structure. (HM: HACCP Module.)

6.3 Describe the products and their intended use

At this stage a product description may be constructed for two reasons. Firstly, it is essential that the HACCP Team is fully familiarized with the products and process technologies to be covered by the HACCP Plan. Secondly, the product description acts as an introduction and point of historical reference to the HACCP Plan (Box 6.2).

Box 6.2 Product description: Iced Delights ice-cream

These are frozen, ready-to-eat products containing both pasteurized and unpasteurized components. The cream, sugar, powders, heat-stable liquids and water are pasteurized, while the flavouring and particulates are added without further heat processing. Air is also whipped into the product at freezing.

The products will be consumed, without further processing, by the general population, including high-risk groups.

The products are packed in containers of 1.0 litre, 500 ml and 100 ml single-serve sizes.

Products included:

Single flavours: vanilla, chocolate, strawberry

Varieties:

(a) **Vanilla base**
Chocolate Chip
Mint Choc Chip
Strawberry Delight
Raspberry Shortbread
Caramel Crunch

(b) **Chocolate base**
Double Chocolate Flake
White Choc Chip
Choc Choc Chip

Each product has undergone an individual product safety assessment covering the formulation, and specific ingredient and process issues.

Principal raw material types:

- dry powders, e.g. skimmed milk powder, cocoa, granulated sugar
- bulk liquids, e.g. cream, liquid sugar
- dry particulates, e.g. chocolate chips, biscuit pieces
- frozen fruit pulps, purées and concentrates
- ambient liquids and pastes, e.g. flavourings, lethicin
- packaging

Key process technologies:

- blending
- homogenization
- pasteurization
- freezing

Key hazards for consideration:

- pathogens in raw materials added post pasteurization
- spore outgrowth at ageing stage
- allergen control
- fruit stalks
- metal
- packaging migration issues
- antibiotics in cream and skimmed milk powder
- gluten in shortbread

Key control measures:

- process steps
- supplier quality assurance activities
- labelling
- cross-contamination prevention
- temperature control

6.4 Constructing a Process Flow Diagram

6.4.1 Types of data

The Process Flow Diagram is used as the basis of the hazard analysis and must therefore contain sufficient technical detail for the study to progress. It should be carefully constructed by members of the HACCP Team as an accurate representation of the process, and should cover all stages from raw materials to end product, as defined in the HACCP Study terms of reference. The following types of data should be included:

- Details of all raw materials and product packaging, including format on receipt and necessary storage conditions.
- Details of all process activities, including the potential for any delay stages. It is important that this lists all the individual activities rather than becoming a list of process equipment.
- Temperature and time profile for all stages. This will be particularly important when analysing microbiological hazards as it is vital to assess the potential for any pathogens present to grow to hazardous levels.
- Types of equipment and design features. Are there any dead-leg areas where product might build up and/or which are difficult to clean?

- Details of any product reworking or recycling loops.
- Floor plan with details of segregated areas and personnel routing. While it is possible to indicate process flow and floor plan on the same diagram, HACCP Teams often find it helpful to keep these as two distinct diagrams in the HACCP Plan.
- Storage conditions, including location, time and temperature.
- Distribution/customer issues (if included in your terms of reference).

6.4.2 Style

The style of the Process Flow Diagram is the choice of each organization and there are no set rules for presentation. However, it is often felt that diagrams consisting solely of words and lines are the easiest to construct and use. Engineering drawings and technical symbols are used by some companies but, because of their complexity, these may cause confusion and so are not advised.

Whichever style of presentation is chosen, a key point is to ensure that every single stage is covered and in the correct order. For large, complex processes, where the modular approach is being used, it is often simplest to prepare a separate diagram for each operation. Where this is done, it is important to show exactly how each diagram fits together and the HACCP Team must ensure that no stages have been missed out, particularly transfer stages and rework.

6.4.3 Verify during manufacture

When the Process Flow Diagram is complete (Figure 6.3) it must be verified by the HACCP Team prior to the hazard assessment stage. This involves team members watching the process in action to make sure that what happens is the same as what is written down, and may also involve going in on the night shift or weekend shift to ensure that any alternatives are included. It is essential to establish that you have got it right as the hazard analysis and all decisions about CCPs are based on these data.

Figure 6.3 outlines the Process Flow Diagrams for our modular HACCP Plan example.

6.5 Carrying out the hazard analysis

When the Process Flow Diagram has been completed and verified, the HACCP Team can move on to the next stage of the HACCP Study, the hazard analysis, as described by HACCP Principle 1. This is one of the key stages in any HACCP Study as the team must

HM2: Non-bulk ingredients – Receipt and Storage

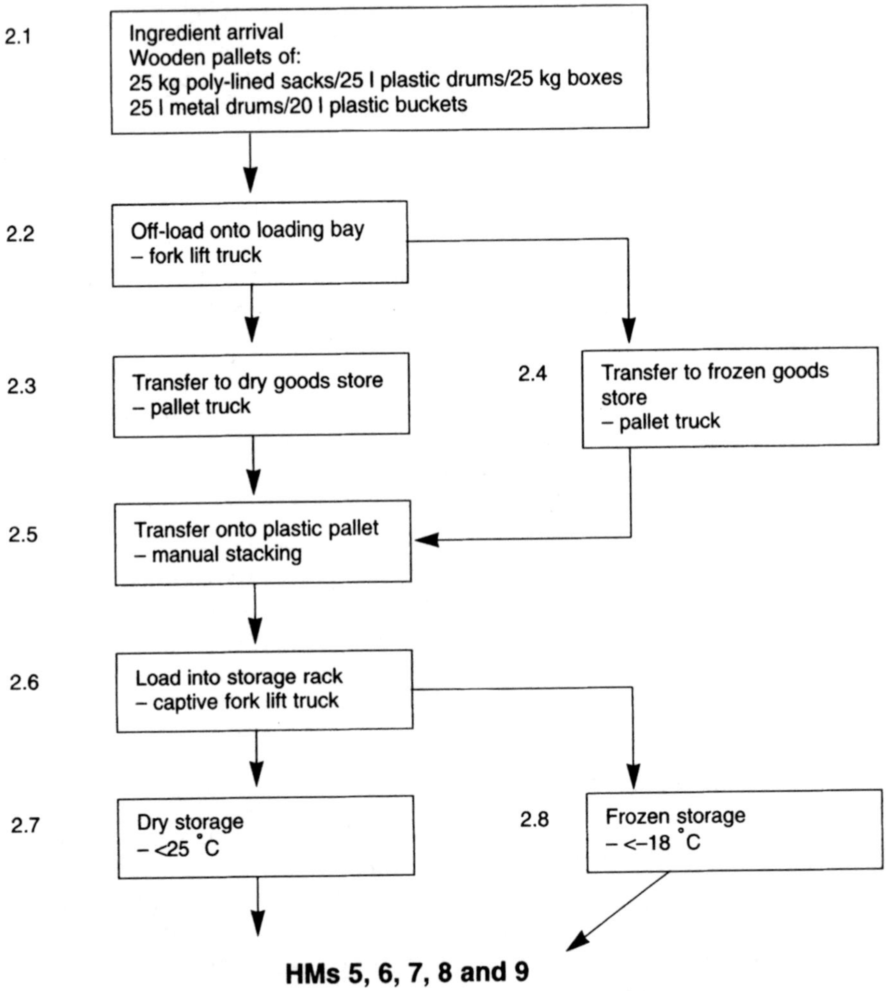

2.1 — Ingredient arrival
Wooden pallets of:
25 kg poly-lined sacks/25 l plastic drums/25 kg boxes
25 l metal drums/20 l plastic buckets

2.2 — Off-load onto loading bay
– fork lift truck

2.3 — Transfer to dry goods store
– pallet truck

2.4 — Transfer to frozen goods store
– pallet truck

2.5 — Transfer onto plastic pallet
– manual stacking

2.6 — Load into storage rack
– captive fork lift truck

2.7 — Dry storage
– <25 °C

2.8 — Frozen storage
– <–18 °C

HMs 5, 6, 7, 8 and 9

Verified: ...... *A. Jones* ....
HACCP Team Leader

Date: .... *5–6–97* ....

Figure 6.3 Process Flow Diagrams – ice-cream manufacture.

HM5: Dry Powder Preparation

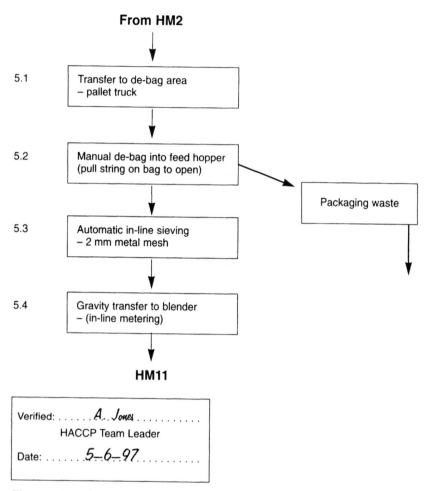

From HM2

5.1 — Transfer to de-bag area – pallet truck

5.2 — Manual de-bag into feed hopper (pull string on bag to open)

Packaging waste

5.3 — Automatic in-line sieving – 2 mm metal mesh

5.4 — Gravity transfer to blender – (in-line metering)

HM11

Verified: . . . . . . *A. Jones* . . . .
HACCP Team Leader

Date: . . . . *5–6–97* . . . .

Figure 6.3 *continued*

ensure that all potential hazards are identified and considered. Hazard analysis involves the collection and evaluation of information on hazards and conditions leading to their presence, in order to decide which are significant for food safety and therefore should be addressed in the HACCP Plan. Several resources and techniques are available to your HACCP Team to assist in this task, as described in the following sections. However, before starting out on the hazard analysis, all team members must be clear on the meaning of the word 'hazard'. Remember, a 'hazard' is normally considered to be a factor which may cause a food to be unsafe for consumption. Hazards can be of biological, chemical or physical nature (Chapter 4).

HM8: Dry Particulates Preparation

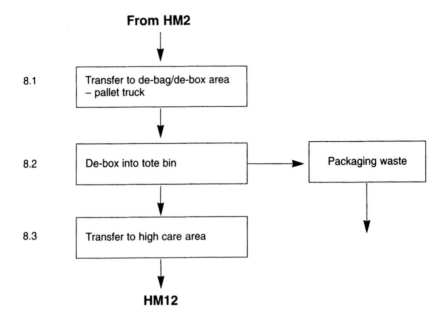

Figure 6.3 *continued*

> **HAZARD:**
> **A biological, chemical or physical property, or condition of, food with the potential to cause an adverse health effect.**

6.5.1 *The structured approach to hazard analysis*

A structured approach to hazard analysis helps to ensure that all conceivable hazards have been identified. It really is crucial that you do not miss any hazards and this will be helped by having personnel from a wide range of disciplines in your HACCP Team, working from a verified Process Flow Diagram.

When your HACCP Team is new to hazard analysis, it is usual to ensure that all potential hazards are identified before moving on to

HM9: Frozen Fruit/Fruit Purée Preparation

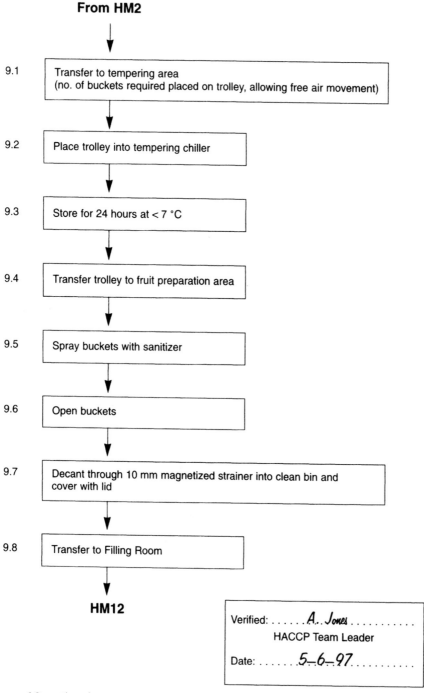

From HM2

9.1 — Transfer to tempering area
(no. of buckets required placed on trolley, allowing free air movement)

9.2 — Place trolley into tempering chiller

9.3 — Store for 24 hours at < 7 °C

9.4 — Transfer trolley to fruit preparation area

9.5 — Spray buckets with sanitizer

9.6 — Open buckets

9.7 — Decant through 10 mm magnetized strainer into clean bin and
cover with lid

9.8 — Transfer to Filling Room

HM12

Verified: . . . . . . *A. Jones* . . . .
HACCP Team Leader

Date: . . . . *5–6–97* . . . .

Figure 6.3 *continued*

137

HM11: Ice-cream Base Manufacture

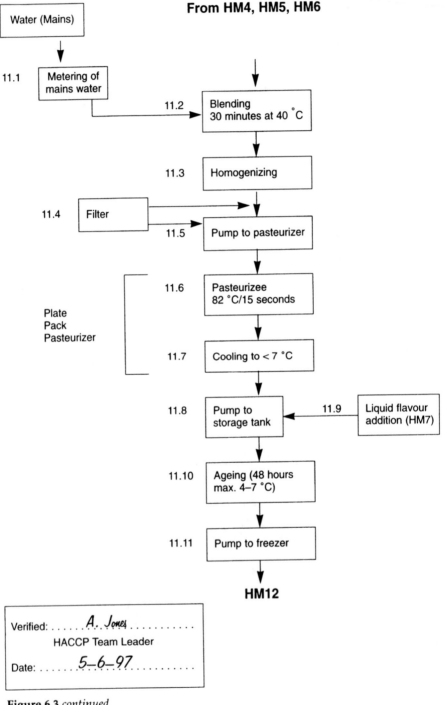

Figure 6.3 *continued*

HM12: Filling Room (ambient temperature < 10 °C)

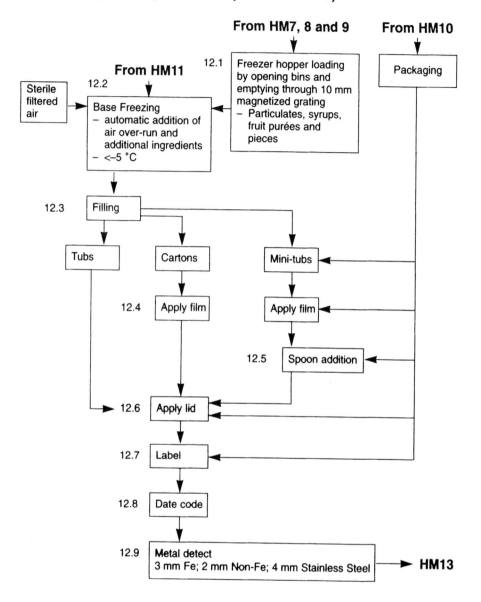

Verified: ...... *A. Jones* ....

        HACCP Team Leader

Date: ...... *5–6–97* ....

Figure 6.3 *continued*

HM13: Low Care Finishing and Storage

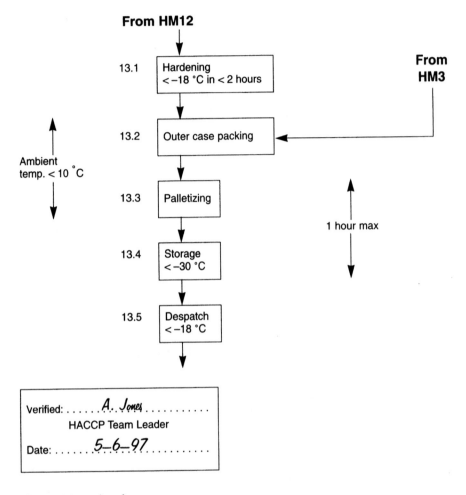

Figure 6.3 *continued*

discuss risk and possible control measures. More experienced teams may wish to discuss control measures already in place or required, at the same time as hazard analysis, as it may save time. But – some words of caution – make sure that you do not miss any hazards out. It is often found that personnel get into deep discussions about the merits of different control measures and that the hazard identification stage either loses momentum or loses its way completely. When this happens hazards may easily be missed, so it can be better to ensure you have identified them all first.

Some organizations have found it helpful to record all hazards as they are identified, against the process steps where they occur, in a

Table 6.1 Hazard Analysis Chart

Process step	Hazard and source/cause	Control measure

structured manner. It is also useful to document the source or cause of each hazard, as this helps in identifying the correct control measure. The documentation produced is then used as the basis for the hazard analysis and discussion of control measures. The use of such informal documentation helps to structure the thinking and discussions of the HACCP Team, and therefore helps to ensure that all potential hazards are covered. An example of such a Hazard Analysis Chart is shown in Table 6.1.

At each stage in the Process Flow Diagram, the hazards and their causes or sources should be brainstormed. This can be done either formally, through a structured brainstorming session, or informally, as part of a general discussion. Brainstorming is one of a number of standard problem-solving techniques that can be applied successfully to HACCP and is particularly useful at the hazard identification step for a number of reasons:

1. Analytical thinking stifles creativity. Where team members are analytically or scientifically trained, lateral thinking and new ideas may be repressed.

2. The group is too close to the process and how it has always been done. This makes it difficult to challenge what is known or understood, and leads to assumptions being made and beliefs being accepted.

3. The belief that there is always one correct solution to every problem. This leads individuals into searching for the one correct answer, and in doing so overlooking alternative, less apparent solutions.

Table 6.2 Ice-cream, freezing stage – hazard brainstorming

Hazard	Cause
Metal	Blades from bag-opening knives
Jewellery	Through de-bag operators not following personal hygiene policy
Crawling insects Condensation Rivets Rust }	Open top on choc chip hopper allows entry of debris from overhead structure
Flying insects	Insects flying into hopper
Polythene	Poor operating practice at de-bagging
Pathogenic micro-organisms in air	Filter malfunction
Cross-contamination • Microbiological • Chemical	Poor freezer cleaning procedures

In order to overcome these barriers, brainstorming is an approach where each team member offers an idea. An individual is usually allocated the position of scribe to ensure that all ideas are recorded and a time limit may be set to keep the pressure on. Brainstorming is often carried out as a facilitated quick-fire session and team members can say whichever hazards come into their heads. It is successful because other team members are able to think laterally and build on the ideas previously suggested. Ideas are never praised, criticized or commented on during the brainstorming session because this may influence contributions about to be made.

As an example, at the freezing stage of the Process Flow Diagram for ice-cream, which includes the addition of the particulates such as chocolate chips, the Iced Delights HACCP Team identified the hazards and their causes shown in Table 6.2 through brainstorming.

Following the brainstorming session, the HACCP Team should analyse all the hazard ideas but must be careful that no idea is rejected, unless all team members are confident that there is no risk in the process under study.

6.5.2 *What is the risk?*

'Would you rather eat mouldy grain and face the chronic low-probability of risk of liver cancer, or would you rather eat nothing at all and face an acute prospect of starvation?' (William H. Sperber, 1995)

Within the HACCP study we need to take a logical, practical approach to risk assessment. At the end of the hazard identification step, the HACCP Team will have a list of potential hazards that might occur in the raw materials or during the process. Risk assessment involves the evaluation of the potential hazards on this list, to establish the realistic or significant hazards that the HACCP system must control. Some useful definitions are:

RISK:
The probability or likelihood that an adverse health effect will be realized.

SIGNIFICANT HAZARD:
A hazard that is likely to occur and which would cause an adverse health effect.

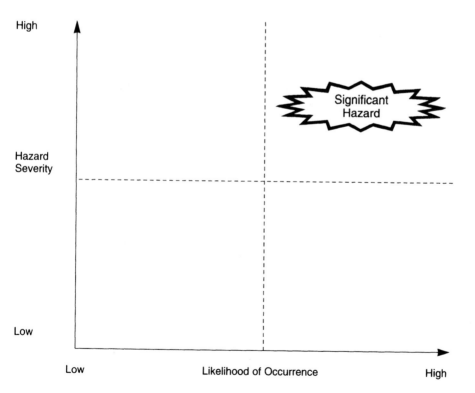

Figure 6.4 Determination of hazard significance.

In other words, a significant hazard is one with a high likelihood of occurrence and severe outcome, as shown in Figure 6.4. You may also wish to consider the likelihood of detection should the control measure fail.

(a) Assessment of risk

Following on from the identification of hazards, the process of risk assessment involves three additional steps (Bernard and Scott, 1995; Sperber, 1995):

- exposure assessment;
- hazard characterization;
- risk characterization.

Exposure assessment is the evaluation of the degree of intake likely to occur when a product containing the hazard is consumed. This can be difficult to establish, particularly for microbiological hazards, where numbers may be affected by handling practices after production in perishable products, or where the organism is irregularly distributed, i.e. non-homogeneous.

Bernard and Scott (1995) list factors that influence levels of micro-organisms in foods as follows:

- Physical treatments, such as hot-holding or freezing, may kill a proportion of the microbial population.
- Refrigeration will allow slow growth of some micro-organisms, including a few pathogens.
- Holding of many foods at temperatures within the range 5–60 °C will allow microbial growth.
- Under certain conditions, spoilage organisms may compete with pathogens, reducing their growth and the likelihood of infection.
- Thorough cooking of foods before consumption will kill vegetative cells of pathogenic organisms, thus virtually eliminating the threat of certain types of infection.

Hazard characterization is an evaluation of the nature of the adverse effects or severity associated with the hazard, and may involve a dose–response assessment, i.e. the effect relative to the amount or numbers of the hazard consumed.

Following exposure assessment and hazard characterization (which may be carried out simultaneously), **risk characterization** is the estimation of the adverse effects likely to occur in the population, i.e. the nature and magnitude of human risk. This is normally carried out at governmental level rather than in individual food businesses.

In practical terms, risk assessment involves expert judgement as to the degree of harm or adverse health effects which may be caused by an uncontrolled hazard in a food product. In other words, the process is identifying the significant hazards which must be controlled by the HACCP System.

Several different approaches to risk assessment can be adopted by the HACCP Team. They include qualitative, quantitative and semi-quantitative techniques.

(b) Risk assessment – quantitative versus qualitative?

Quantitative risk assessment techniques are often used in the field of chemical toxicity. These are detailed studies based on knowledge of likely total exposure to specific chemicals in the food chain, and often experimental effect data, e.g. from studies on mice or rats, are extrapolated to potential effects on humans.

Quantitative risk assessment has also been carried out for microbiological hazards, where historical information on likely microbial or toxin intake in illness outbreaks has been used to estimate the potential adverse effects for particular micro-organisms in foods. This is really at best semi-quantitative risk assessment, since the number of different factors affecting human susceptibility to infections makes precise evaluation difficult. Notermans *et al.* (1995) summarized the findings of the ILSI-Europe working group of infectious dose, as shown in Table 6.3.

Because of the problems associated with obtaining meaningful data, quantitative risk assessment is not normally the first choice of the HACCP Team. Instead, qualitative risk assessment is more commonly used. This involves forming a judgement on risk, based on knowledge of the product/process and likely occurrence/severity of the hazard, and understanding of the likely use or abuse of the

Table 6.3 Some factors affecting human susceptibility to foodborne infections (adapted from Notermans *et al.*, 1995)

Host Factors	Food Factors	Organism Factors
Age	Fat content	Dose
Pregnancy	Metal content	Virulence
Nutritional status	Acidity	
Concurrent or recent infections	Background flora	
Physiological factors (gastric acidity, gastric transit time, gut flora, antibiotics)	Preservatives	
Stress	Temperature	
	Irradiation	
	Physical state (liquid/solid)	
	Buffering	
	Storage history	

Table 6.4 Structured risk assessment

Hazard	Likelihood of occurrence			Severity of outcome			Comment
	High	Medium	Low	High	Medium	Low	
For example, presence of *Salmonella* in raw chicken	✓			✓			Severe outcome if not controlled by the cooking process

product before consumption. It should be stressed that, to form this judgement correctly, the HACCP Team must include an appropriate level of experience and expertise. Inexperienced HACCP Teams must recognize their limitations and bring in additional resource where necessary.

Some practitioners like to use a structured approach to the assessment, identifying high, medium and low risk. An example of this is shown in Table 6.4. This could be incorporated into the Hazards Analysis Chart shown in section 6.5.1 (Table 6.1) to form a hazard analysis and risk assessment chart or table.

Other HACCP teams prefer to carry out risk assessment as a team discussion, noting at the end their findings on which hazards have been considered realistic and significant.

Whichever approach is taken, a number of questions will need to be asked about the potential hazards and about the facility and its processes and systems. Some initial questions to help prompt discussions can be found in the next section.

6.5.3 Questions to be considered

The following includes some hazard evaluation questions based on those put together by the NACMCF (1997); however, the list is not exhaustive, and you may have some additional ideas.

Remember to refer to your individual product information and the product description, particularly where using the modular approach.

(a) Prerequisites

Are prerequisite programmes in place? What do they cover? Can they be shown to have effectively designed any hazards out of the process? Are adequate verification procedures in place to ensure that they are working?

146

(b) Raw materials

What hazards are likely to be present in each raw material and are these likely to be of concern to the process and/or product? Are any of the raw materials themselves hazardous if excess amounts are added?

(c) Design of plant and equipment

Where are there risks of cross-contamination occurring during the process and at any holding stages? Consider microbiological, chemical and physical safety issues. Are there any stages where contamination could build up or where micro-organisms might grow to dangerous levels?

Can the equipment be effectively controlled within the required tolerances for safe food production? Can effective cleaning be carried out? Are there any extra hazards associated with particular equipment?

(d) Intrinsic factors

Do the product's integral factors (pH, a_w, etc.) effectively control all microbiological hazards likely to be present in the raw materials or which could enter the product as cross-contaminants during the process? Remember there are different types of micro-organisms which react in different ways – what will control one might not control another. Which intrinsic factors must be controlled to ensure product safety? Will microbiological hazards survive or grow in the product formulation?

(e) Process design

Will microbiological hazards survive any heating step in the process or is there a step which will destroy all pathogens? Does the use of reworked or recycled product during the process or in any of the raw materials cause a potential hazard? Consider carefully both microbiological hazards and their toxins. Is there a risk of recontamination between process stages?

(f) Facility design

Are there any hazards directly associated with the facility layout or internal environment? Is segregation adequate between raw and ready-to-eat product? Is positive-pressured filtered air necessary? Do movement patterns for personnel and equipment cause any hazards?

(g) Personnel

Could personnel practices affect the safety of the product? Are all food handlers adequately trained in food hygiene? Are occupational health procedures in place? Do all employees understand the aims and significance of the HACCP System, along with where their role affects the process?

(h) Packaging

How does the packaging environment influence the growth and/or survival of microbiological hazards? For example, is it aerobic or anaerobic? Does the package have all required labelling and instructions for safe handling and use, and can these be easily understood? Is the package damage-resistant and are tamper-evident features in place where required? Is coding sufficient to allow product traceability and recall?

(i) Storage and distribution – what could go wrong?

Could the product be stored at the wrong temperature and will this affect safety during the shelf-life? Could the product be abused by the customer causing it to be unsafe? Can the product be traced through the distribution chain in a timely manner, and withdrawn from the market place in the event of a food safety issue?

(j) Customer complaints

Are there any identifiable trends in customer complaint data? Does this suggest uncontrolled or inadequately controlled hazards in the process?

6.5.4 Reference materials – where to find them and how to use them

A wealth of reference material is available to assist you in identifying and analysing the hazards in your process.

The members of your HACCP Team are the first important resource with their collective and multi-disciplinary experience of the process and product technology under study. Different team members will be able to contribute, for example, an indication of what hazards are likely to be found in the raw materials, where contamination could occur or build up in the process, what the process is capable of, etc. The team as a whole will be able to discuss the significance of these individual issues and ascertain the risk or likelihood of each hazard occurring.

In areas where the team expertise is limited it is important to know where information and advice can be obtained. It is essential that further expertise is secured when required as incorrect evaluation and predictions could have food safety implications.

Chapter 4 gives detail on a wide range of biological, chemical and physical hazards, along with control measure options. Although this list of hazards and control measures is not exhaustive, it can be a good starting point for the HACCP Team and can be used to spark off other ideas during a hazard brainstorming session.

Examples of hazards in different product and raw material types can easily be found in the literature in the form of books, epidemiological reports and research papers. Here again information can be found on the likely behaviour of particular hazard types during the process along with possible controlling options. When using literature data to assist with hazard analysis it is vital that the HACCP Team is able to interpret this data and evaluate the significance to the process under study. Information may also be found through the Internet, in hazard databases and through the use of mathematical models. Legislation may also help where it highlights particular concerns with specific product types. Again, it is important to be able to interpret the significance of any data found.

If you do not have sufficient expertise available there are a number of organizations and resources where you may obtain support. These include industry bodies, research associations, higher educational establishments, regulatory enforcement authorities and external expert consultants, also your suppliers of raw materials and customers.

6.5.5 *Identifying control measures*

When all potential hazards have been identified and analysed, the HACCP Team should go on to list the associated control measures. These are the control mechanisms (also known as preventative measures) for each significant hazard and are normally defined as those factors that are required in order to prevent, eliminate or reduce the occurrence of hazards to an acceptable level. Examples of control measures can be found in Chapter 4.

CONTROL MEASURES:
Any factor or activity which can be used to prevent, eliminate or reduce to an acceptable level, a food safety hazard.

When evaluating control measures it is necessary to consider what you already have in place and what new measures may need to be put in place. This can easily be done using your Process Flow Diagram and/or Hazard Analysis Charts as a guide, and taking into account the source or cause of the hazard.

Table 6.5 Hazard Analysis Chart

Ice-cream manufacture HACCP Module 2 (HM2)
Non-bulk ingredients – receipt and storage

Process step	Hazard and source	Control measure	Authors' notes
2.1 Ingredient arrival	No hazard identified	–	
2.2 Offload onto loading bay	Chemical contamination: migration of exhaust fumes through packaging. High-fat materials and biscuit crumbs susceptible	Vehicle engines to be switched off on docking. Introduce new new third-party drivers' code to site security rules	This is a taint issue and is unlikely to be truly hazardous. However, a detectable off-flavour may cause the consumer to feel unwell and perceive a greater risk. The additional control measures identified here would become part of the ongoing GMP programme
2.3 Transfer to dry goods store	No hazard identified	–	
2.4 Transfer to frozen goods store	No hazard identified	–	
2.5 Transfer onto plastic pallet	No hazard identified	–	
2.6 Load into storage rack	No hazard identified	–	
2.7 Dry storage	No hazard identified	–	

			However, temperature abuse may lead to fruit texture deterioration
2.8 Frozen storage	No hazard identified	–	

HM5, 6, 7, 8 and 9

Ice-cream manufacture HACCP Module 5 (HM5) Dry powder preparation

Process step	Hazard and source	Control measure	Authors' notes
From HM2			
5.1 Transfer to de-bag area	No hazard identified	–	
5.2 Manual de-bag into feed hopper (pull string on bag to open and rip)	Foreign material – string from bag and large packaging fragments	GMP programme In-line sieving	Referring back to the definition of a hazard (a property which may cause food to be unsafe for consumption), these would not normally be perceived as significant hazards and, with an adequate prerequisite GMP programme, would not be likely to occur. However, food safety legislation in the UK (Food Safety Act, 1990) goes beyond true hazards and covers all foreign material
5.3 Automatic in-line sieving	Foreign material carried through from raw materials/de-bagging due to sieve malfunction	Planned preventative maintenance – part of GMP programme	

Table 6.5 *continued*

Ice-cream manufacture HACCP Module 5 (HM5)
Dry powder preparation

Process step	Hazard and source	Control measure	Authors' notes
5.4 Gravity transfer to blender → **HM11**	No hazard identified	–	

Ice-cream manufacture HACCP Module 8 (HM8)
Dry particulates preparation

Process step	Hazard and source	Control measure	Authors' notes
From HM2 →			
8.1 Transfer to de-bag/de-box area	No hazard identified	–	
8.2 Transfer into tote bin	Large packaging fragments	GMP programme	See previous note at step 5.2
8.3 Transfer to high-care area → **HM12**	No hazard identified	–	

Ice-cream manufacture HACCP Module 9 (HM9)
Frozen fruit/purée preparation

Process step	Hazard and source	Control measure	Authors' notes
From HM2			
9.1 Transfer to tempering area	No hazard identified	—	
9.2 Place trolley into tempering chiller	No hazard identified	—	
9.3 Store for 24 hours at < 7 °C	No hazard identified	—	
9.4 Transfer trolley to fruit preparation area	No hazard identified	—	
9.5 Spray buckets with sanitizer	No hazard identified	—	Cross-contamination of the fruit/purée with sanitizer is not considered to be a hazard due to the small quantities used and the low risk of transfer right through the bucket, even if damaged
9.6 Open buckets	No hazard identified	—	
9.7 Decant through 10 mm magnetized strainer into clean bin and cover with lid	Metal carry-through due to magnet malfunction (e.g. magnet not cleaned regularly)	Planned preventative maintenance: magnet pull strength calibration and cleaning programme	

Table 6.5 *continued*

Process step	Hazard and source	Control measure	Authors' notes
	Foreign material carry-through due to strainer damage	Planned preventative maintenance	
9.8 Transfer to filling room	No hazard identified		
↓			
HM12			

Ice-cream manufacture HACCP Module 11 (HM11)
Ice-cream base manufacture

Process step	Hazard and source	Control measure	Authors' notes
From HM4, 5 and 6			
↓			
11.1 Metering of mains water	No hazard identified	–	* The possibility of product contamination with high levels of pathogens/toxins should be considered, depending on equipment design and cleanability. Here, an effective prerequisite cleaning programme is in place, verified before each use by ATP hygiene monitoring of surfaces and rinse waters
11.2 Blending of ingredients	No hazard identified*	–	
11.3 Homogenizing	No hazard identified*	–	

11.4 In-line filtration	Carry-through of foreign material due to filter malfunction	Planned preventative maintenance – filter in place and intact	The likelihood of foreign material being present and of a size which is injurious to health is considered to be remote. Therefore this is not a true hazard
11.5 Pump into pasteurizer	No hazard identified*	–	
11.6 Pasteurize ⎫ plate pack ⎬	Survival of vegetative pathogens through not achieving correct heat process (time and/or temperature)	Correct heat process achieved	
11.7 Cooling ⎭	Cross-contamination with pathogens – leakage from raw side of plate pack due to damage and/or inadequate pressure differential	Correct pressure set up; planned preventative maintenance	
11.8 Pump to storage tank	No hazard identified*	–	
11.9 Add liquid flavour	No hazard identified	–	
11.10 Ageing	Outgrowth of spore-forming pathogens due to temperature abuse	Temperature maintenance < 7 °C Maximum storage 48 hours	

Table 6.5 *continued*

Ice-cream manufacture HACCP Module 11 (HM11)
Ice-cream base manufacture

Process step	Hazard and source	Control measure	Authors' notes
11.11 Pump to freezer ⟶ **HM12**	No hazard identified*		

Ice-cream manufacture HACCP Module 12 (HM12)
Filling room

Process step	Hazard and source	Control measure	Authors' note
From HM7, 8, 9, 10 and 11 ⟶			
12.1 Freezer hopper loading – particulates, syrups, fruit purées and pieces – by opening bins through magnetized grating into freezer hopper	Metal carry-through due to magnet malfunction	Planned preventative maintenance – magnet pull strength calibration and cleaning programme	Some ingredients, e.g. dry particulates, have not been through a magnetized grating prior to this step
	Foreign material carry-through due to grating damage	Planned preventative maintenance	

12.2 Base freezing – automatic addition of sterile filtered air over-run and particulates/syrups	Introduction of pathogens in contaminated air	Effective filtration and planned preventative maintenance	The possibility of product contamination through poor cleaning of the freezer should also be considered here
12.3 Fill tubs/cartons	Hazardous foreign material ingress from environment or filling heads	Planned preventative maintenance and GMP	With effective prerequisite GMP and preventative maintenance the risk is small
12.4 Apply film (cartons/minitubs only)	No hazard identified	–	
12.5 Spoon addition (minitubs only)	No hazard identified	–	
12.6 Apply lid	No hazard identified	–	
12.7 Label	No hazard identified	–	There are no major allergens within the present range, therefore the possibility of wrongly labelling the product presents a legal non-compliance risk rather than safety

Table 6.5 *continued*

Ice-cream manufacture HACCP Module 12 (HM12)
Filling room

Process step	Hazard and source	Control measure	Authors' notes
12.8 Date code	No hazard identified other than inability to trace and recall product	Correct date coding	There is not strictly a food safety hazard here. However, in order to minimize the effect if a CCP fails, it is vital to be able to trace and recall all products concerned. Therefore it may be appropriate to manage this as a CCP. Additionally, lot traceability is likely to be a legal requirement
12.9 Metal detect → HM13	Metal contamination not identified due to equipment malfunction	Effective metal detection – calibrated metal detector suitable for product dimensions Planned preventative maintenance	

Ice-cream manufacture HACCP Module 13 (HM13)
Low-care finishing and storage

Process step	Hazard and source	Control measure	Authors' notes
From HM3, 12 →			
13.1 Hardening	No hazard identified	–	Temperature abuse during low-care finishing and storage would result in textural quality problems rather than safety hazards
13.2 Outer case packing	No hazard identified	–	
13.3 Palletizing	No hazard identified	–	
13.4 Storage < –30 °C	No hazard identified	–	
13.5 Despatch < –18 °C	No hazard identified	–	

Remember that more than one control measure may be required to control a hazard which occurs at different stages of the process. For example, the potential for contamination with *Listeria monocytogenes* before and after cooking in a ready-to-eat product. For contamination before cooking the heat process might be the control measure, while environmental control would be required to prevent contamination after cooking. Similarly, more than one hazard might be effectively controlled by one control measure, e.g. two microbiological pathogens by a heat process, or glass and metal by sifting.

The completed Hazard Analysis Chart shows the hazards and control measures identified by the HACCP Team for the modules being studied in our ice-cream example (Table 6.5).

6.6 Where are the Critical Control Points?

HACCP Principle 2 requires that Critical Control Points (CCPs) are identified.

6.6.1 How to find them

As we saw in Chapter 4, a CCP is a point, step or procedure where a food safety hazard can be prevented, eliminated or reduced to acceptable levels.

CCP:
A step where control can be applied and is essential to prevent, eliminate or reduce a food safety hazard to acceptable levels.

CCPs can be found by using your thorough knowledge of the process and all the possible hazards and measures for their control. The information established during the hazard analysis should allow the identification of CCPs through the expert judgement of the HACCP Team and specialist advisers.

However, the location of CCPs using judgement alone may lead to disagreement within the HACCP Team, and may result in more points being managed as CCPs than are really necessary. There is always the tendency to err on the side of caution, but designating too many points as CCPs, rather than correctly identifying the real CCPs, may mean that you lose credibility and commitment as there will always be some points where you are prepared to negotiate a deviation. For example, if a metal detector failed at a raw material stage, you could switch it off and rely on the one at the end of the line, which, as long as the sensitivity is appropriate, would lead to the consequence of potential substandard stock which is later rejected, and not compromised food safety.

On the other hand, too few CCPs would be even more disastrous and could cause the sale of unsafe food. It is important that control is focused where it is essential for food safety and so care should be employed to ensure that the CCPs are correctly identified.

To assist in finding where the correct CCPs should be, a tool is available known as the CCP Decision Tree (Figure 6.5). As we saw in Chapter 5, a decision tree is a logical series of questions that are asked for each hazard. In the case of the CCP Decision Tree this is for each hazard at each process step. The answer to each question leads the HACCP Team through a particular path in the tree and to a decision whether or not a CCP is required at that step.

Using a CCP Decision Tree promotes structured thinking and ensures a consistent approach at every process step and for each hazard identified. It also has the benefit of forcing and facilitating team discussion, further enhancing teamwork and the HACCP Study.

Several versions of the CCP Decision Tree have been published (CCFRA, 1992, 1997; NACMCF, 1992, 1997; Codex, 1993, 1997) with slightly different wording, although they display a common approach to CCP location. Figure 6.5 has been adapted from the above in order to simplify the approach.

6.6.2 Use of CCP Decision Trees

The questions in the tree should be asked for each hazard at each process step, including receipt and handling of raw materials.

If you have used the previous decision tree for raw materials at the sourcing or Supplier Quality Assurance stage (see Chapter 5) you will already know where specific CCPs are required to control incoming raw materials before they reach your site. If you were to put the same raw materials through the questions in the CCP Decision Tree you would find the same answers.

However, you will find that it helps to focus on raw materials first at the development stage, and should pick these up again with the CCP Decision Tree where they arrive as incoming goods at your facility.

Work through the CCP Decision Tree as follows:

Q1 Is there a hazard at this process step?
This first question will seem obvious but it helps to focus the HACCP Team's minds on the specific process step in question. This is an additional question to that published within Codex (1997) and is particularly useful if there is a time delay between carrying out the hazard analysis and determining CCPs. Sometimes a 'hazard' identified during brainstorming turns out not to be a real hazard when challenged here; perhaps it is a mistakenly identified quality issue. If there is a hazard then you should move on to Q2.

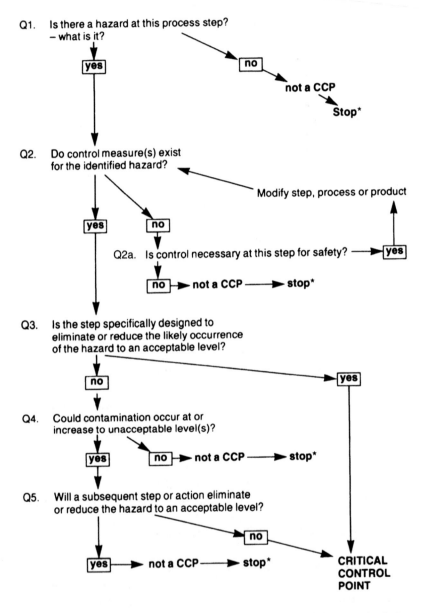

Q1. Is there a hazard at this process step?
– what is it?

yes

no

not a CCP

Stop*

Q2. Do control measure(s) exist
for the identified hazard?

Modify step, process or product

yes

no

Q2a. Is control necessary at this step for safety? ⟶ yes

no ⟶ not a CCP ⟶ stop*

Q3. Is the step specifically designed to
eliminate or reduce the likely occurrence
of the hazard to an acceptable level?

no

yes

Q4. Could contamination occur at or
increase to unacceptable level(s)?

yes

no ⟶ not a CCP ⟶ stop*

Q5. Will a subsequent step or action eliminate
or reduce the hazard to an acceptable level?

no

yes ⟶ not a CCP ⟶ stop*

CRITICAL
CONTROL
POINT

* Stop and proceed with the next hazard at the current step or the next step in the
described process

Figure 6.5 A CCP Decision Tree (adapted from Codex, 1997).

162

Q2 Do control measures exist for the identified hazard?
Here you need to consider the measures you already have in place along with what could be implemented, and this is most easily done using your Hazard Analysis Charts. If the answer to this question is yes, then you should move straight on to Question 3.

If, however, the answer is no and control measures are not and could not be put in place, then you must consider whether control is necessary at this point for food safety (Q2a). If control is not necessary here then a CCP is not required and you should move on to the next hazard and start the decision tree again. However, make sure that if you are answering no here because there is control later on, that you actually pick up the later point as a CCP. An example of how this question loop works is for metal detection. Metal detection might not be required for safety at some of the early process steps although they may be associated with a metal hazard. It would, of course, be essential to have a metal detector at the finished-product stage.

If members of the HACCP Team have identified a hazard at a process step and there are no possible control measures at that or any following step, then you must carry out a modification to build in control. This may involve either the process step, the process itself, the product, or the introduction of a new procedure such that food safety control is possible. For example, if *Salmonella* is likely to be present and your heat process is not sufficient to destroy the organism, then you will need to look at increasing your heat process or building in some other control method. It should be noted that a process step can only operate as a CCP if control measures can be introduced. When the necessary modifications have been established you should ask Q2 again and progress through the tree.

It is important that any necessary changes highlighted at this stage by the HACCP Team, to the step, process, product or procedures, are agreed and implemented before the product goes into production. Here you may need to ensure that senior management fully accept the HACCP Team's findings, and provide the required back-up for the change(s) to be implemented.

Q3 Is the step specifically designed to eliminate or reduce the likely occurrence of the hazard to an acceptable level?
The key thing to remember when asking this question is that it is the process step and not the control measure that is being questioned. If team members incorrectly consider the control measure, then they will always answer yes, and additional, unnecessary CCPs will result. This question was originally developed to accommodate process steps which are specifically designed to control specific hazards. What the question is really asking is whether the step itself

controls the hazard. For example, milk pasteurization at 71.7 °C for 15 seconds is specifically designed and will control vegetative pathogens, while ambient storage of raw materials is not specifically designed to control hazards such as pest infestation.

You should consider carefully your hazard analysis information along with the Process Flow Diagram to answer this question, and remember – it is just as important to consider mixing steps, where it is critical to get the product formulation right, as it is to consider the main processing steps. If the product is not properly mixed, then your intrinsic control mechanisms may not be effective, and an unmixed product may have detrimental effects on other processing steps, e.g. the heat process.

If the answer is yes, then the process step in question is a CCP and you should start the decision tree again for the next process step or hazard. If the answer is no, move on to Q4.

Where there is debate over whether or not the process step is 'specifically designed' to control the hazard, it is worth noting that an alternative route through the decision tree should give the same ultimate answer. For example, if the process step is chilled storage and the hazard is microbial (pathogen) growth, it could be argued that the step is not specifically designed to control the hazard, but to have perishable material readily available for the next step of process/distribution. In this case, two possible routes through the decision tree apply but the outcome is the same:

Q1	Q2	Q2a	Q3	Q4	Q5	CCP
Y	Y	–	Y	–	–	Yes
Y	Y	–	N	Y	N	Yes

Q4 Could contamination occur at or increase to unacceptable level(s)?

This question requires your hazard analysis information along with the HACCP Team's combined experience of the process and processing environment. The answer should be largely obvious from the hazard analysis but make sure that you have covered the following issues:

- Is the immediate environment likely to include the hazard(s)?
- Is cross-contamination possible via personnel?
- Is cross-contamination possible from another product or raw material?
- Could composite time/temperature conditions increase the hazard?
- Could product build up in dead-leg spaces and increase the hazard?

• Are any other factors or conditions present which could cause contamination to increase to unacceptable levels at this step?

Where there is uncertainty about what constitutes unacceptable levels of a particular factor (i.e. where it becomes a hazard), it is important that the HACCP Team should seek expert advice before making a decision. However, if a completely new process is under study, it may not be possible to obtain a definite answer. Here the HACCP Team should always assume that the answer is yes and proceed appropriately.

When considering how contamination could increase to unacceptable levels, it is important to understand the possible additive effect during the process for each particular factor. This means that you may need to think not only about the current process step, but also whether any subsequent steps or holding stages between steps could cause the hazard to increase. For example, a number of steps being performed at ambient temperature might give the opportunity for a low initial contamination level of *Staphylococcus aureus* to grow to toxin-forming levels, and become a hazard.

If the answer to Q4 is yes, i.e. contamination could occur at or increase to unacceptable levels, move on to the next question. If the answer is no, go back to the beginning of the decision tree with the next hazard or process step.

Q5 Will a subsequent step or action eliminate or reduce the hazard to an acceptable level?

This question is designed to allow the presence of a hazard or hazards at a particular process stage if they will be controlled either later in the process or by consumer action. In this way it minimizes the number of process steps which are considered to be CCPs and focuses on those steps which are crucial for product safety.

If the answer to this question is yes, then the current process step is not a CCP for the hazard under discussion but the subsequent step/action will be. For example, correct consumer cooking will control some of the microbiological hazards present in a raw meat product. Similarly, metal detection of finished products at the packing stage will detect metal contamination which may be a hazard associated with the raw materials or an earlier process stage. If the answer is no, then the current process step must be a CCP for the hazard being considered.

Although this question allows the number of CCPs to be minimized, this may not be appropriate in all cases. In the above example of metal detection, the only CCP which is absolutely critical is metal detection at the finished product stage. However, from a commercial point of view, the early detection/control of metal or any other hazard where there is a high degree of risk will be advisable and additional preventative control points (see section 4.3) may be built in. When this is done it must be made clear that the purpose is to establish additional control in order to minimize product losses.

There may also be occasions where the cost of the control measure itself is prohibitive. An example of this may be the X-ray detection of bone in meat, where there may only be one X-ray detector available. Here it must be located at the CCP, as this takes priority over any additional manufacturing controls.

All CCPs identified by the HACCP Team must be implemented and cannot be replaced by other measures elsewhere in the process.

When working through the decision tree you may find it helpful to designate members of the HACCP Team as questionmaster and scribe. The team will find it much easier to vocalize hazards and process steps as they progress through the decision tree, e.g. at Q3: 'Is the (insert process step under consideration) step specifically designed to eliminate or reduce the likely occurrence of (insert hazard under consideration) to an acceptable level?' This will ensure that the discussions are structured and that the team does not become side-tracked. It is often also helpful to construct a question-and-answer matrix for each process step where a hazard has been identified. This could be done on a flip chart so that all team members can see it. Alternatively, this could be an extension to the Hazard Analysis Chart.

You should continue to work through the decision tree for each hazard present at each process step until all CCPs have been determined. When this has been achieved the HACCP Team should highlight the CCPs on the Process Flow Diagram and move on to building up the HACCP Control Chart.

Before moving on to look at how CCPs are controlled, we will follow through the CCP Decision Tree for the example ice-cream HACCP modules (Table 6.6).

6.7 Building up the HACCP Control Chart

As we saw in section 6.1, the HACCP Control Chart is one of the key documents in the HACCP Plan, holding all the essential details about the steps or stages in the process where there are CCPs. This information could be documented separately elsewhere, but most companies find it easier to hold it all together in one matrix, such as the example shown in Table 6.7.

6.7.1 What are the Critical Limits?

When you have identified all the CCPs in your process, the next step is to decide on their safety boundaries. This is HACCP

Table 6.6 Process step decision matrix – ice-cream manufacture

Process step – hazard	Q1	Q2	Q2a	Q3	Q4	Q5	CCP	HACCP Team notes
HACCP Module 2: non-bulk ingredients receipt and storage								
2.2 Offload onto loading bay – taint from exhaust fumes	N	–		–	–	–	No	This was recognized as a quality defect. Driver's procedures prevent the occurrence
HACCP Module 5: dry powder preparation								
5.2 Manual de-dag into feed hopper – string and large packaging fragments	N	–		–	–	–	No	On further discussion, the team decided that these would not be hazardous to health, and should be prevented by adherence to GMP
5.3 Automatic in-line sieving – foreign material due to sieve malfunction	N	–		–	–	–	No	Powders are also pre-sieved at supplier (SQA). Not a hazard at this step but preventative maintenance will act as an additional preventative control point
HACCP Module 9: frozen fruit/purée preparation								
9.7 Decant through 10 mm magnetized strainer into clean bin and cover with lid – metal carry-through due to magnet malfunction	Y	Y	–	N	Y	Y	No	Metal detection later on in process

Table 6.6 Process step decision matrix – ice-cream manufacture

Process step – hazard	Q1	Q2	Q2a	Q3	Q4	Q5	CCP	HACCP Team notes
– foreign material carry-through due to strainer damage	N	–	–	–	–	–	No	The team considered that it was unlikely that such a robust strainer would be damaged
HACCP Module 11: ice-cream base manufacture								
11.4 In-line filtration – carry-through of foreign material due to filter malfunction	N	–	–	–	–	–	No	Health risk unlikely
11.6 Pasteurize – survival of vegetative pathogens	Y	Y	–	Y	–	–	Yes	Pasteurization is specifically designed to kill vegetative pathogens
11.7 Cooling – cross-contamination with pathogens	Y	Y	–	N	Y	N	Yes	This type of hazard is commonly associated with plate-pack pasteurizers
11.10 Ageing – outgrowth of spore-forming pathogens	Y	Y	–	N	Y	N	Yes	Temperature control is critical in prohibiting spore germination and outgrowth

HACCP Module 12: filling room

	Q1	Q2	Q3	Q4	Q5	Q6	CCP?	
12.1 Freezer hopper loading – metal carry-through due to magnet malfunction	Y	Y	–	N	Y	Y	No	Metal detection later in process
– foreign material carry-through due to grating damage	N	–	–	–	–	–	No	Robust grating similar to strainer in process step 9.7
12.2 Base freezing – introduction of pathogens in contaminated air	N	–	–	–	–	–	No	High-efficiency microbiological filters in place as part of GMP
12.3 Fill tubs/cartons – hazardous material ingress from environment or filling heads	N	–	–	–	–	–	No	Small risk and therefore not considered a significant hazard
12.8 Date code – inability to trace and recall product	Y	Y	–	Y	–	–	Yes	The HACCP team have decided that, in order to minimize risk of illness or injury caused through inability to trace and recall, the step will be managed as a CCP
12.9 Metal detect – metal contamination due to equipment malfunction	Y	Y	–	Y	–	–	Yes	This is the final opportunity for control of metal before distribution and sale to the consumer

Principle 3. You must establish the criteria that indicate the difference between safe and unsafe product being produced so that the process can be managed within safe levels. The absolute tolerance at a CCP, i.e. the division between safe and unsafe, is known as the Critical Limit. If the Critical Limits are exceeded, then the CCP is out of control and a hazard may exist.

CRITICAL LIMITS:
Criteria that separate acceptability from unacceptability.

The product will be safe as long as all the CCPs are managed within their specified Critical Limits. Critical Limits may meet (or exceed) national/international regulations, company safety standards or scientifically proven values.

(a) How do you set the Critical Limits?

Since the Critical Limits define the boundaries between safe and unsafe product, it is vital that they are set at the correct level for each criteria. The HACCP Team must therefore fully understand the criteria governing safety at each CCP in order to set the appropriate Critical Limit. In other words, you must have detailed knowledge of the potential hazards, along with a full understanding of the factors that are involved in their prevention or control. Critical Limits will not necessarily be the same as your existing processing parameters.

Each CCP may have a number of different factors which need to be controlled to ensure product safety, and each of these factors will have an associated Critical Limit. For example, cooking has long been established as a CCP that destroys vegetative pathogens. Here the factors associated with control are temperature and time. The Critical Limits associated with industrial meat cooking are that the centre temperature achieves 70 °C for at least 2 minutes. In cases such as this, where numerical Critical Limits are required the limit will be an absolute value and not a range.

In order to set the Critical Limits, all the factors associated with safety at the CCP must be identified. The level at which each factor becomes the boundary between safe and unsafe is then the Critical Limit. It is important to note that the Critical Limit must be associated with a measurable factor that can be monitored routinely by test or observation. Some factors that are commonly used as Critical Limits include temperature, time, pH, moisture or a_w, salt concentration and titratable acidity.

As HACCP Team members you will have an in-depth knowledge of the hazards and control mechanisms of the process, and you may have an understanding of the safety boundaries.

Table 6.7 The HACCP Control Chart

CCP no.	Process Step	Hazard	Control Measure	Critical Limits	Monitoring			Corrective action	
					Procedure	Frequency	Responsibility	Procedure	Responsibility

However, in a number of cases this may be beyond your in-house expertise and it is again important to know where you can obtain information and advice. Possible sources of information are as follows:

- Published data – information in scientific literature, the Internet, in-house and supplier records, industry and regulatory guidelines (e.g. Codex, ICMSF, FDA, IDF).
- Expert advice – from consultants, research associations, plant and equipment manufacturers, cleaning chemical suppliers, microbiologists, toxicologists and process engineers.
- Experimental data – these are likely to support Critical Limits for microbiological hazards and may come from planned experiments, challenge studies where product is inoculated, or from specific microbiological examination of the product and its ingredients.
- Mathematical modelling – computer simulation of the survival and growth characteristics of microbiological hazards in food systems.

(b) Types of Critical Limit

The factors or criteria that make up the Critical Limit will be related to the type of hazard that the CCP is designed to control, and the specific control measure in place. They may involve numbers – either minimum or maximum values for the given criteria but never a range – for example, maximum pH 4.5 to prevent growth of *Listeria monocytogenes*, minimum temperature/time combination for HTST milk pasteurization 71.7 °C for 15 seconds.

• Chemical limits

These may be associated with the occurrence of chemical hazards in the product and its ingredients or with the control of microbiological hazards through the product formulation and intrinsic factors. Examples of factors involved in chemical limits are maximum acceptable levels for mycotoxins, pH, salt and a_w, or the labelling or absence of allergens.

• Physical limits

These are often associated with the tolerance for physical or foreign material hazards. However, they can also be involved in the control of microbiological hazards, where the survival or death of the micro-organism is governed by physical parameters. Examples of factors associated with physical limits are absence of metal, intact sieve (sieve size and retention), and temperature and time.

• **Procedural limits**

These will be associated with procedural control measures. For example, 'debag procedure followed at all times' may be a Critical Limit where the control measure to prevent packaging materials entering the product stream is a specific debag procedure. Similarly, the limit might be 'continued approved status' where the control measure is effective supplier assurance for particular hazards. Where these control measures are part of prerequisite programmes, it would be **critical** that the prerequisite elements were working correctly.

• **Microbiological limits**

These should normally be avoided as part of the HACCP System, apart from the control of non-perishable raw materials. This is because microbiological factors can usually only be monitored by growing the organism of concern in the laboratory, a process which may take several days. The monitoring of microbiological limits would therefore not allow you to take instant action when the process deviates. Instead you might have several days' production quarantined in storage, without knowing where the hazard is present. This is further complicated by the fact that micro-organisms are rarely distributed homogeneously throughout a batch, and therefore may be completely missed (remember the limitations of inspection and testing as discussed in Chapter 2). It may be possible to use microbiological limits for positive release of raw materials, but only if the material is homogeneous and a representative, statistically valid sample can be taken.

Microbiological factors are best kept for verification purposes, i.e. where you perform additional tests to ensure that the HACCP System has been effective, as here the time scale involved does not create operational difficulties. One exception to this general rule is where rapid microbiological methods can be implemented, but even these need to be truly rapid, i.e. minutes rather than hours, to be effective. An example here is ATP bioluminescence, which can be used to demonstrate the effectiveness of cleaning procedures, and the polymerase chain reaction techniques, which may be used more widely in the future for a number of applications.

When your HACCP Team have established appropriate Critical Limits for all CCPs, they should be added to the HACCP Control Chart as in the example in Table 6.8.

In addition to your Critical Limits you may find it beneficial to have another layer of control to help you manage the process. This can be done by setting up target or action levels within your Critical

Table 6.8 Ice-cream manufacture – HACCP Control Chart

HACCP Plan No. HP001 Page 1 of 3				Iced Delights HACCP Control Chart					Date Approved by HACCP Team Leader	Supersedes		
CCP No.	Process Step	Hazard	Control Measure	Critical Limits	Monitoring					Corrective Action		
					Procedure	Frequency	Responsibility			Procedure	Responsibility	
	HACCP Module Ingredients:											
I.1	Skimmed milk powder	Antibiotic residues	Effective supplier assurance – Audit	Continued approved status (Audit pass)								
			Certificate of Analysis – Agreed specification (maximum acceptable levels)	Legal limits								

I.2	Cream	Antibiotic residues	Effective supplier assurance – Audit	Continued approved status (Audit pass)
			Certificate of Analysis – Agreed specification (maximum acceptable levels)	Legal limits
I.3	Chocolate chips	Salmonella	Effective supplier assurance: – Audit	Continued approved status (Audit pass)

Table 6.8 *continued*

HACCP Plan No. HP001 Page 2 of 3			Iced Delights HACCP Control Chart			Date Supersedes Approved by . . . . HACCP Team Leader					
						Monitoring			Corrective Action		
CCP No.	Process Step	Hazard	Control Measure	Critical Limits		Procedure	Frequency	Responsibility	Procedure	Responsibility	
			– Agreed specification (maximum acceptable levels)	Absent/ 50 g							
I.4	Plastic tubs and film	Chemical – plasticizers and additives – leach into product	Correct choice of container and film (agree in specification)	Suitable for food use: – High fat product – Compliance with legal migration limits							
			Effective supplier assurance – Audit	Continued approved status (Audit pass)							

	HACCP Module 11: Ice-cream base manufacture			
11.1	Pasteurizing	Survival of vegetative pathogens	– Correct heat process, target level 82 °C/15 seconds	79.4 °C/15 seconds
			– Automtic divert	Divert working
11.2	Cooling	Cross-contamination with pathogens due to leaks in plate pack	– Correct pressure differential	Constant pressure
11.3	Ageing	Spore outgrowth due to poor temperature control and batch stock rotation	Effective temperature control	7 °C maximum
			Effective stock rotation	48 hours maximum

Table 6.8 *continued*

Iced Delights
HACCP Control Chart

Date   Supersedes 12–8–96
Approved by ....
HACCP Team Leader

CCP No.	Process Step	Hazard	Control Measure	Critical Limits	Monitoring			Corrective Action		
					Procedure	Frequency	Responsibility	Procedure	Responsibility	
	HACCP Module 12: Filling room									
12.1	Date coding	Inability to trace and recall product resulting in unfit product in marketplace	Effective date and batch coding	Correct code applied						
12.2	Metal detecting	Metal in packed product	Effective metal detection	2.0 mm ferrous, 3.0 mm non-ferrous, 4.0 mm stainless steel						

Limits. The target levels can be used as an additional measure to indicate drift in the process, and you can then adjust the process to maintain control before the CCP actually deviates from its Critical Limits. In other words a buffer zone for safety is provided (Figure 6.6). An example of target levels can be found at the pasteurization step during ice-cream production. The Critical Limits for the destruction of vegetative pathogens through the heat process are 79.4 °C for 15 seconds. In order to make sure that deviation does not occur, the process parameters might be set at 82 °C for 15 seconds, the target levels.

TARGET LEVELS:
Control criteria which are more stringent than Critical Limits, and which can be used to take action and reduce the risk of a deviation.

Operating your system to target levels should ensure that a deviation from the Critical Limits never occurs. They are set for day-to-day management of the process. Target levels may be added to the HACCP Control Chart, e.g. in the Control Measures column, or you may choose to document them elsewhere. You will, however, need to ensure that they are documented and that they tie in with your monitoring procedures. The best way of doing this is to document the target levels on your monitoring log sheets, and you must ensure that all monitoring personnel understand how they work.

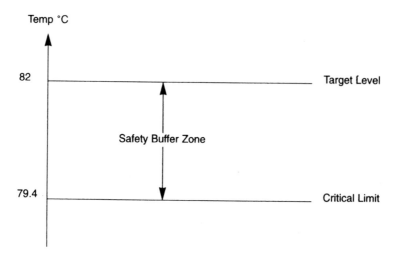

Figure 6.6 Critical Limits and target levels for the example ice-cream heat process.

6.7.2 Finding the right monitoring procedure

Monitoring is the measurement or observation at a CCP that the process is operating within the Critical Limits. This is HACCP Principle 4. It is one of the most important parts of the HACCP System, ensuring that the product is manufactured safely from day to day.

MONITOR:
The act of conducting a planned sequence of observations or measurements of control parameters to assess whether a CCP is under control.

The specific monitoring procedure for each individual CCP will depend on the Critical Limits, and also on the capabilities of the monitoring device or method. It is essential that the chosen monitoring procedure must be able to detect loss of control at the CCP (i.e. where the CCP has deviated from its Critical Limits), as it is on the basis of monitoring results that decisions are made and action is taken.

Monitoring procedures may involve:

- On-line systems, where the critical factors are measured during the process. These may be **continuous** systems where critical data are continuously recorded, or **discontinuous** systems where observations are made at specified time intervals during the process.
- Off-line systems, where samples are taken for measurement of the critical factors elsewhere. Off-line monitoring is normally discontinuous (unless a continuous sampling device has been used) and this has the disadvantage that the sample taken may not be fully representative of the whole batch.
- Observational procedures, where a specific action is observed by the monitor. For example, observation by a supervisor that the operator is following the debag procedure correctly.

Most monitoring systems are based on some form of inspection and testing. We pointed out the limitations of inspection and testing as control measures in Chapter 2 and, although these procedures do have serious limitations as control measures, they can be useful monitors that control has been achieved (by the control measures at the CCPs). As monitoring procedures, inspection and testing activities are properly targeted on critical factors throughout the process, and set up through statistical means in order to demonstrate ongoing control. Increasingly, automated on-line sensors are being used for monitoring which increase confidence that control is being achieved.

The frequency of monitoring will depend on the nature of the

CCP and the type of monitoring procedure. It is imperative that an appropriate frequency is determined for each monitoring procedure. For example, in the case of a calibrated metal detector, the frequency of checks is likely to be every 30 minutes, while with a seasonal vegetable crop, the CCP for pesticides may be monitored by pesticide testing once per crop season.

It is vital that inspection and testing programmes being used to monitor CCPs are statistically valid, and these programmes will be of most benefit if established under a structured system of Statistical Process Control. This will be covered in Chapter 7.

(a) Who should be responsible?

The most important issue with responsibility is that you ensure that it is properly defined. All personnel involved need to understand clearly what they are required to do, and also how to do it. These details should be decided by the HACCP Team in conjunction with other management and must be fully documented on the HACCP Control Chart.

As we have seen, monitoring is a key part of the HACCP System operation and it is therefore vital that the persons involved in monitoring understand and are fully accountable for their monitoring actions. Monitoring procedures are closely related to the production process, so it is usually most appropriate that the responsibility for monitoring lies with the Production Department.

We can now fill in the monitoring procedures and frequencies, along with responsibility for each activity, in our HACCP Control Chart example, shown in Table 6.9.

6.7.3 Corrective action requirements

HACCP Principle 5 requires that corrective action be taken when the monitoring results show a deviation from the Critical Limit(s) at a CCP. However, since the main reason for implementing HACCP is to prevent problems from happening in the first place, you should also endeavour to build in corrective actions which will prevent deviation from happening at the CCP. Your HACCP Plan is therefore likely to have two levels of corrective action, i.e. actions to prevent deviation and actions to correct following deviation.

Corrective action procedures should be developed by the HACCP Team and should be specified on the HACCP Control Chart. This will minimize any confusion or disagreements which might otherwise have occurred when the action needs to be taken.

Table 6.9 Ice-cream manufacture – HACCP Control Chart

HACCP Plan No. HP001 Page 1 of 3			Iced Delights HACCP Control Chart					Date Approved by HACCP Team Leader	Supersedes		Corrective Action	
						Monitoring						
CCP No.	Process Step	Hazard	Control Measure	Critical Limits	Procedure	Frequency	Responsibility			Procedure	Responsibility
	HACCP Module Ingredients:										
I.1	Skimmed milk powder	Antibiotic residues	Effective supplier assurance – Audit	Continued approved status (Audit pass)	– Audit by trained SQA auditor	Annual	SQA Auditor				
			Certificate of Analysis – Agreed specification (maximum acceptable levels)	Legal limits	– Check Certificate of Analysis	Each delivery	Incoming Goods Clerk				

I.2	Cream residues	Antibiotic supplier assurance – Audit	Effective			
			Continued approved status (Audit pass)	– Audit by trained SQA auditor	Annual	SQA auditor
		Certificate of Analysis – Agreed specification (maximum acceptable levels)	Legal limits	– Check Certificate of Analysis	Each delivery	Incoming Goods Clerk
I.3	Chocolate chips	Salmonella	Effective supplier assurance: – Audit			
			Continued approved status (Audit pass)	Audit by trained SQA auditor	Annual	SQA Auditor
		– Agreed specification (maximum acceptable levels)	Absent/ 50 g	Check supplier certificate of analysis for compliance	Every batch	Goods Inwards Clerk

Table 6.9 *continued*

Iced Delights
HACCP Control Chart

Date .... Supersedes ....
Approved by ....
HACCP Team Leader

CCP No.	Process Step	Hazard	Control Measure	Critical Limits	Monitoring			Corrective Action	
					Procedure	Frequency	Responsibility	Procedure	Responsibility
I.4	Plastic tubs and film	Chemical – plasticizers and additives – leach into product	Correct choice of container and film (agree in specification)	Suitable for food use: – High fat product – Compliance with legal migration limits	Review component listing and supplier migration data against legislation	Every time pack type or packaging supplier changes	QA Manager		
			Effective supplier assurance – Audit	Continued approved status (Audit pass)	Audit by trained SQA auditor	Annual	SQA Auditor		
	HACCP Module 11: Ice-cream base manu-facture								

11.1	Pasteurizing	Survival of vegetative pathogens	– Correct heat process, target level 72 °C/15 seconds	79.4°C/15 seconds	Chart recorder – Visual inspection and sign off	Each batch	Production Operator
			– Automatic divert	Divert working	Check automatic divert operation	Twice daily (start and end of production)	Production Operator
					Check temperature sensor against traceable calibrated thermometer	Daily	Production Operator
11.2	Cooling	Cross-contamination with pathogens due to leaks in plate pack	– Correct pressure differential	Constant pressure	Check pressure gauge	Daily start-up	Production Operator

Table 6.9 *continued*

HACCP Plan No. HP001 Page 3 of 3				Iced Delights HACCP Control Chart					Date Supersedes Approved by . . . . HACCP Team Leader			
							Monitoring			Corrective Action		
CCP No.	Process Step	Hazard	Control Measure	Critical Limits		Procedure	Frequency	Responsibility	Procedure	Responsibility
11.3	Ageing	Spore outgrowth due to poor temperature control and batch stock rotation	Effective temperature control	7 °C maximum		Chart recorder – Visual inspection and sign off	Each shift	Production Operator		
						Check temperature sensor against traceable calibrated thermometer	Every batch	QA Technician		
			Effective stock rotation	48 hours maximum		Record date and time in and out of ageing tank	Daily	Production Operator		

	HACCP Module 12: Filling room					
12.1	Date coding	Inability to trace and recall product resulting in unfit product in marketplace	Effective date and batch coding	Correct code applied	Visual inspection	Start-up and half-hourly	Production Operator
12.2	Metal detecting	Metal in packed product	Effective metal detection	2.0 mm ferrous, 3.0 mm non-ferrous, 4.0 mm stainless steel	Check metal detector with test pieces	Start-up and half-hourly Include end of run	Production Operator

It is also important to assign responsibility for corrective action both to prevent and correct deviations. This will be discussed in more detail at the end of this section.

CORRECTIVE ACTION:
Action to be taken when the results of monitoring at the CCP indicate a trend towards or actual loss of control.

As we have seen, there are two main types of corrective action.

1. Actions which adjust the process to maintain control and prevent a deviation at the CCP

This first type of corrective action normally involves the use of target levels within the Critical Limits. When the process drifts towards or exceeds the target levels it is adjusted, bringing it back within the normal operating bands.

This is typified by on-line continuous monitoring systems which automatically adjust the process, e.g. automatic divert valves in milk pasteurization which open when the temperature falls below the target level, sending milk back to the unpasteurized side. However, preventative corrective action can also be associated with manual monitoring systems where the CCP monitor takes action when the target levels are approached or exceeded, and thus prevents a CCP deviation.

The factors that are often adjusted to maintain control include temperature and/or time, pH/acidity, ingredient concentrations, flow rates and sanitizer concentrations. Some examples are as follows:

- Continue to cook for longer to achieve the correct centre temperature.
- Add more acid to achieve the correct pH.
- Chill rapidly to correct storage temperature.
- Add more salt to the recipe.

When adjusting the process to maintain control, you must ensure that you can do so without causing or increasing the hazard. For example, if the product temperature had risen above 5 °C and you implement rapid chilling to bring it back down, then you must know that the temperature has not risen high enough for long enough to allow the growth of any microbiological hazards which might be present.

2. Actions to be taken following a deviation at a CCP

Following a deviation it is important to act quickly. You will need to take two types of action and it is vital that detailed records are kept.

- Adjust the process to bring it back under control.

    This may involve stopping and restarting the line if it is not possible to return the process to its normal operating level during production. Possibly a corrective action will involve the provision of a short-term repair so that production can restart quickly with no more deviations, while the permanent corrective action takes a longer period of time, e.g. the provision of temporary off-line metal detection until the in-line metal detector is repaired.
- Deal with the material which was produced during the deviation period.

    In order to handle non-complying materials effectively you will need to implement a series of further actions:

(1) Place all suspect product on hold.
(2) Assess the situation, seeking advice from the HACCP Team, facility management and other relevant experts. Here it is important to consider the likelihood of the hazard being present in the product.
(3) Conduct further tests, where appropriate, to assess safety.

When you have obtained sufficient information, the decision about what should happen to the product can be taken. This would probably be to:

(1) destroy the non-complying product;
(2) rework into new products;
(3) direct non-complying product into less sensitive products such as animal feed; or
(4) release product following statistically based sampling and testing.

Destruction of the non-complying product is the most obvious action, and the main one to be taken when the likelihood of the hazard occurring, in products which cannot be reworked, is high. However, this has the disadvantage of being costly, and is therefore normally the action of last resort.

Reworking the product can be carried out where the hazard would be controlled through the reworking process. It is important to ensure that any reworking does not cause new hazards in the secondary product, e.g. when allergenic ingredients such as nuts are reworked into a product where they will not appear on the pack ingredients listing. The key here is to rework like with like.

If the product can be diverted into another safe use then this is another option. This might involve packing as less-sensitive animal feeds, or possibly diverting into another product where the hazard will be controlled, e.g. the use of cooked meats, contaminated with vegetative micro-organisms, in pie fillings which will receive another cook. Here the presence of heat-stable toxins must be carefully considered. Again, the presence of allergenic materials must be controlled.

You may decide to sample and test the product to establish whether or not the hazard is present. As previously discussed, great care must be taken when implementing sampling regimes due to the statistical probability of detecting the hazard. Do you know what the probability of finding the hazard is using your chosen sampling plan, and if so, are you confident that the remaining product will be safe?

A final option is simply to release the product, but this cannot ever be advised for safety. You have chosen to implement HACCP to prevent safety problems and are designing your HACCP Plan to control hazards. This is what your CCPs are set up for, and this action shows a lack of commitment to the HACCP concept. Product safety is not negotiable and product manufactured during a deviation cannot simply be released. The company's legal position, if hazardous products were knowingly sold, also needs to be considered.

It is important that detailed records are kept of all stages. It is essential that you investigate the cause of the deviation, and take appropriate steps to ensure that it does not happen again. The defined corrective action procedures are added to the HACCP Control Chart which should detail:

- what is to happen to the suspect product;
- how the process/equipment is to be adjusted;
- who is to do what;
- who is to be informed.

Responsibility for corrective action

Responsibility for corrective action will again often lie with the Production Department who are implementing the HACCP Plan, but you should consider assigning particular responsibilities at different levels in the management structure.

On-line responsibilities of the CCP monitor or line operator will often involve the notification of a supervisor who will then co-ordinate further actions. However, responsibility may also be given at this level for stopping the line or adjusting the process in order to prevent large quantities of product being made while the CCP is out of control.

Off-line, more senior responsibility will be appropriate where the corrective actions involve shutting down the plant for periods of time or where disposition actions are required. These decisions need to be taken by personnel who have the knowledge to recommend the appropriate corrective action for product manufactured during a deviation, as outlined in the previous section. This may involve the HACCP Team Leader in discussion with facility management. However, if the HACCP Team Leader is an expert in HACCP techniques rather than in hazards and their associated risks, it is important that other experts should be involved in the decision-making process, e.g. toxicologists, microbiologists, process specialists.

It is also important to ensure that the individuals who are responsible for documenting and signing off the corrective action procedures are defined. This information will be crucial in proving that the required action has been taken, particularly important for legal issues.

At this stage the HACCP Control Chart should be complete as in our example in Table 6.10.

The locations of CCPs should then be added on the final Process Flow Diagram for retention in the HACCP Plan, as in Figure 6.7.

6.8 Process capability – can the Critical Limits be met?

As part of the HACCP Plan the Critical Limit for each Critical Control Point within the process has been established. These limits may sometimes only be a minimum value, such as the time and temperature requirements for a heat treatment process, or the limits may be solely a maximum value, such as cold storage temperature. Other CCPs may require a process to be contained between a minimum and a maximum limit, e.g. nitrite in bacon, where the minimum level controls microbiological safety but the maximum level is necessary to ensure chemical safety. Alternatively, it may also be necessary to have a minimum limit in terms of food safety, but also to have a maximum limit in terms of product quality.

For each CCP you will need to verify that, under normal operating conditions, the process can be realistically and consistently maintained within these defined limits. One way of assessing whether a process is capable is to use statistical analysis. Such statistical techniques have been developed and used for many years, predominantly for process monitoring and control in the engineering industry. The techniques are not really difficult to apply, but for those with no prior knowledge the use of a good reference book or, better still, an expert in the field will be invaluable.

Table 6.10 Ice-cream manufacture – completed HACCP Control Chart

HACCP Plan No. HP001 Page 1 of 4		Iced Delights HACCP Control Chart					Date Approved by HACCP Team Leader	Supersedes		
						Monitoring		Corrective Action		
CCP No.	Process Step	Hazard	Control Measure	Critical Limits	Procedure	Frequency	Responsibility	Procedure	Responsibility	
	HACCP Module Ingredients:									
I.1	Skimmed milk powder	Antibiotic residues	Effective supplier assurance – Audit	Continued approved status (Audit pass)	– Audit by trained SQA auditor	Annual	SQA Auditor	Change supplier	Purchasing Manager	
			Certificate of Analysis – Agreed specification (maximum acceptable levels)	Legal limits	– Check Certificate of Analysis	Each delivery	Incoming Goods Clerk	– Report to QA Manager – Contact supplier – Reject consign-ment	Incoming Goods Clerk QA Manager	

No.	Product	Hazard	Control measure	Critical limit	Monitoring procedure	Frequency	Responsibility	Corrective action	Responsibility
I.2	Cream	Antibiotic residues	Effective supplier assurance – Audit	Continued approved status (Audit pass)	– Audit by trained SQA auditor	Annual	SQA auditor	Change supplier	Purchasing Manager
			Certificate of Analysis – Agreed specification (maximum acceptable levels)	Legal limits	– Check Certificate of Analysis	Each delivery	Incoming Goods Clerk	– Report to QAM – Contact supplier – Reject consignment	Incoming Goods Clerk QA Manager
I.3	Chocolate chips	Salmonella	Effective supplier assurance: – Audit	Continued approved status (Audit pass)	Audit by trained SQA auditor	Annual	SQA Auditor	Change supplier	Purchasing Manager

Table 6.10 *continued*

HACCP Plan No. HP001 Page 2 of 4				Iced Delights HACCP Control Chart					Date Approved by Supersedes .... HACCP Team Leader				
								Monitoring				Corrective Action	
CCP No.	Process Step	Hazard	Control Measure	Critical Limits	Procedure	Frequency	Responsibility				Procedure	Responsibility
			– Agreed specification (maximum acceptable levels)	Absent/ 50 g	Check supplier certificate of analysis for compliance	Every batch	Goods Inwards Clerk				Contact supplier Reject consign-ment	QA Manager
I.4	Plastic tubs and film	Chemical – plasticizers and additives – leach into product	Correct choice of container and film (agree in specifica-tion)	Suitable for food use: – High fat product – Com-pliance with legal migration limits	Review component listing and supplier migration data against legislation	Every time pack type or packaging supplier changes	QA Manager				Change container/ supplier	Purchasing Manager
			Effective supplier assurance – Audit	Continued approved status (Audit pass)	Audit	Annual	SQA Auditor				Change supplier	Purchasing Manager

HACCP Module 11: Ice-cream base manufacture

No.	Process	Hazard	Critical limit	Critical limit	Monitoring	Frequency	Responsibility	Corrective action	Verification
11.1	Pasteurizing	Survival of vegetative pathogens	– Correct heat process, targel level 82 °C/15 seconds	79.4°C/15 seconds	Chart recorder – Visual inspection and sign off	Each batch	Production Operator	– Report to Manager – Contact QA and discuss – Ensure divert working correctly – Repair thermo-graph – Hold product until correct heat process verified; dump if not	QA and Production Managers, Engineer

Iced Delights
HACCP Control Chart

Date Supersedes
Approved by . . . .
HACCP Team Leader

CCP No.	Process Step	Hazard	Control Measure	Critical Limits	Monitoring			Corrective Action	
					Procedure	Frequency	Responsibility	Procedure	Responsibility
			– Automatic divert	Divert working	Check automatic divert operation	Twice daily (start and end of production)	Production Operator	Start: Production does not start until corrected End: Quarantine product; call engineer; notify QA and discuss as above	QA and Production Managers, Engineer
					Check temperature sensor against traceable calibrated thermometer	Daily	Production Operator	– Quarantine product (rework or disposal decision) – Call engineer – Re- calibrate sensor	Operations Manager QA Manager Production Operator Engineer

	Process step	Hazard	Control measure	Critical limit	Monitoring procedure	Frequency	Responsibility	Corrective action	Responsibility
11.2	Cooling	Cross-contamination with pathogens due to leaks in plate pack	– Correct pressure differential	Constant pressure	Check pressure gauge	Daily start-up	Production Operator	– Call engineer; – Notify QA; Production does not start until corrected	Production Operator Engineer
11.3	Ageing	Spore outgrowth due to poor temperature control and batch stock rotation	Effective temperature control	7 °C maximum	Chart recorder – Visual inspection and sign off	Each shift	Production Operator	Quarantine product Contact QA and discuss	Production Operator QA Manager
					Check temperature sensor against traceable calibrated thermometer	Every batch	QA Technician	Quarantine product and discuss with QA	Production Operator
			Effective stock rotation	48 hours maximum	Record date and time in and out of ageing tank	Daily	Production Operator	Quarantine product and discuss with QA	QA Technician

Table 6.10 *continued*

HACCP Plan No. HP001
Page 3 of 4

Iced Delights
HACCP Control Chart

Date   Supersedes
Approved by   . . . .
HACCP Team Leader

CCP No.	Process Step	Hazard	Control Measure	Critical Limits	Monitoring			Corrective Action		
					Procedure	Frequency	Responsibility	Procedure	Responsibility	
	HACCP Module 12: Filling room									
12.1	Date coding	Inability to trace and recall product resulting in unfit product in marketplace	Effective date and batch coding	Correct code applied	Visual inspection	Start-up and half-hourly	Production Operator	Quarantine product and recode Input correct code or repair as appropriate	Production Supervisor/ Operator	
12.2	Metal detecting	Metal in packed product	Effective metal detection	2.0 mm ferrous, 3.0 mm non-ferrous, 4.0 mm stainless steel	Check metal detector with test pieces	Start-up and half-hourly Include end of run	Production Operator	– Repair/ recalibrate metal detector – Quarantine and recheck product back to previous good check	Production Operator QA Supervisor Engineer	

HM11: Ice-cream Base Manufacture

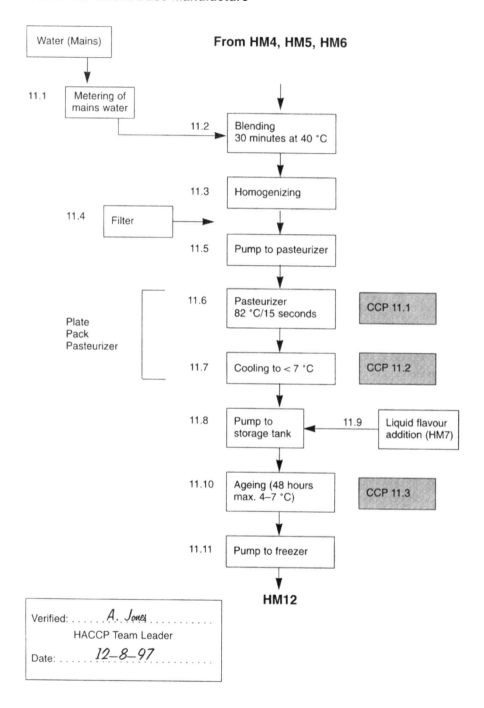

Figure 6.7 Process Flow Diagram, example – showing CCP positioning.

6.8.1 Confirming process capability

The statistical verification of a process in order to establish the probability (confidence) of its ability stay within specified limits is known as establishing the **process capability** (Figure 6.8).

A stable process is one that is statistically in a state of control. In assessing the process capability we are doing two things:

1. determining whether the process is capable of achieving the control criteria (the Critical Limits) that have been established.
2. determining whether the process is capable of being controlled.

All processes are subject to natural and inherent variability. This type of variation is known as 'common cause' variation and is usually the result of a combination of many small sources of variation within the process. If the common cause variation is known, then we know over what range the process is capable of being controlled. Some processes are subject to 'special cause' variation, where the source can be attributed to an unexpected change. These special causes can usually be investigated and corrective action taken to prevent a recurrence.

In establishing the process capability we want to be sure that the process is only subject to common cause variation, i.e. in statistical process control, and that common causes are minimized (Figure 6.9).

There are a few basic requirements for the application of process capability:

1. A series of random samples/readings are taken from the process in consecutive groups of 5–10 (with 50–100 samples taken in total). These measurements of the process must be obtained at a time when all process controls were left untouched throughout the duration of the run.

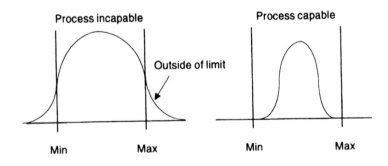

Figure 6.8 Process capability.

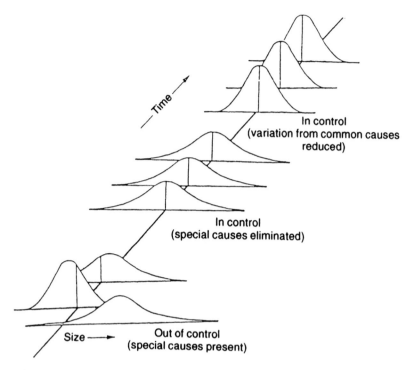

In control
(variation from common causes
reduced)

In control
(special causes eliminated)

Size ⟶   Out of control
(special causes present)

Figure 6.9 Stages of process improvement.

2. These readings must be shown to be normally distributed. A normal distribution must meet mathematically defined requirements, and has a characteristic bell-shaped appearance (Figure 6.10). If this type of distribution can be verified as being contained with the defined limits throughout the process, then the process can be said to be running in 'statistical process control'.
3. The degree of natural variation (spread) of results within the normal distribution can be quantified numerically by statistical analysis. This measurement is known as the **standard deviation (SD)**. From the standard deviation it can be determined by calculation whether the process is capable, or not capable, of running within defined limits.

Standard deviation × 2 = ± variation from the mean value within which 95.4% of the process readings/samples would be expected to be found.

Standard deviation × 3 = ± variation from the mean value within which 99.7% of the process readings/samples would be expected to be found.

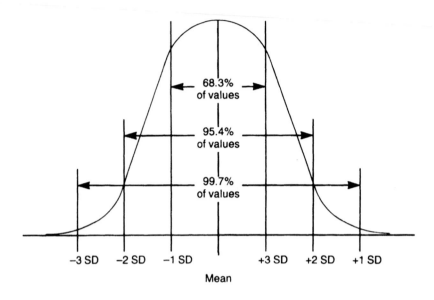

Figure 6.10 The normal distribution.

6.9 Challenging your controls

You are now ready to implement your HACCP Plan and may wish to go straight on to validation. However, before going ahead it is worthwhile pausing to consider whether you really do have sufficient control for all possibilities. What happens, for example, when a CCP fails? Do you have the appropriate contingency plans in place?

It is important to challenge your controls and to understand what would happen in the event of a failure. This can be done using a structured approach considering the consequences of CCP failure.

In order to challenge your CCPs, you will need to investigate all the possible failure modes, the contributory cause(s) of the failure and what the outcome would be. Then consider your current controls and any additional controls required to give more confidence and increase the effectiveness of the HACCP System.

This approach also depends on teamwork and, in particular, the brainstorming approach to identify all the possible failure modes and their associated causes. It requires the team to think the unthinkable and challenge commonly held beliefs such that all potential outcomes are considered. When this has been done it is relatively straightforward to identify and implement an extra 'safety net' of controls so that deviations at CCPs are minimized, and equally important that deviations are not missed through failure in monitoring.

It may be appropriate to use other personnel (e.g. a different

HACCP or Management Team) to carry out this process, as the original HACCP Team may be too close to the CCPs they have identified.

An example of this approach is shown in Table 6.11. The team should use brainstorming to fill in the first three columns, and then discuss whether the CCPs, Critical Limits and monitoring procedures (i.e. the 'current controls') identified during the HACCP Study are sufficient and effective. Following this exercise the final column can be completed as the team considers any extra control criteria required for a more fail-safe system. Some Failure Mode and Effect Analysis (FMEA) methods include a risk-scoring approach (Abbott, 1992) which can be helpful.

The example given shows the considerations for metal hazards being found in the product, but no scoring has been used here.

Table 6.11 Challenging metal detection control

Failure	Outcome of failure	Cause of failure	Current control	Recommended controls
Failure to detect metal	Metal in product	Metal detector breakdown	Metal detect and check metal detector hourly. Record result	Metal detector function verified at start-up.
	– injury	Metal detector not properly calibrated		Maintenance schedule
	– complaints of metal in product f rom customers	Metal detector in wrong place in line		Set up calibration schedule for correct sensitivity
		Incorrect metal detector		
	– lost credibility	Rejects not controlled		Move to just before packing
	– prosecution	Damage in transit		Confirm sensitivity appropriate for all products
	– bad publicity			Locked cage for rejects
	– lost customers			Training for drivers. Work instructions and audits

When you are confident that all your proposed controls are sufficient for all possible outcomes you can move on to validation and implementation of your HACCP Plan.

6.10 Validation of the HACCP Plan

When you have completed your HACCP Control Chart and highlighted all CCPs on your Process Flow Diagram then the HACCP Plan is complete. However, before going on to implement the plan it is important to know that it is correct and valid – a final check that you have got it right. This should be carried out soon after the plan is completed so that implementation can follow without delay.

You should work through all the records in the Process Flow Diagram and HACCP Control Chart to make sure that all the details are actually relevant to the hazards, and that the control criteria, i.e. the Critical Limits, have been set at tight-enough levels to ensure control of product safety. It is equally important that you ensure that no hazards have been missed during the study.

It is also essential to inspect the processing area in order to make sure that all required control measures (particularly new measures) are in place. Critical process and monitoring equipment should also be examined to ensure that it is capable of achieving the desired control criteria and is appropriately calibrated.

Although members of the HACCP Team can carry out some or all of the HACCP Plan validation, it may be appropriate to use other experts to cross-check the study and ensure that no issues have been missed. It is really important to know that you have got it right, so if it is your first HACCP Plan then you should involve other relevant experts in the validation. This could be done by other experts within your company, e.g. at corporate level, or by external independent specialists and HACCP experts.

7

Putting the HACCP Plan into practice

Now that we have seen how to put together a HACCP Plan, the next step is to implement it in your operation. You have made a commitment to use the HACCP System in order to effectively control all safety issues, and unless the HACCP Plan is properly implemented, its real benefits will not be realized. This is a vital stage and yet the relief at having completed the studies can sometimes mean that businesses see the documentation as the end in itself.

In order to implement the system properly, you must ensure that sufficient resources are available, so that the identified CCPs are monitored, and that records are kept. It is also important that the workforce is able to take over the day-to-day running of the system and they may need training and support in order to fully take ownership of HACCP. All of this needs careful planning. In this chapter we will be considering each of these areas and, in addition, how to meet the requirements of HACCP Principle 7.

7.1 Implementation requirements

Several requirements will be involved in implementing the HACCP Plan but there are likely to be several different approaches to implementation. There is no one right or wrong way and the method chosen should reflect the maturity of the existing business and the resources available to it.

All production operations are operating under certain constraints and these are most likely to be associated with time and money. When implementing the HACCP Plan it is important

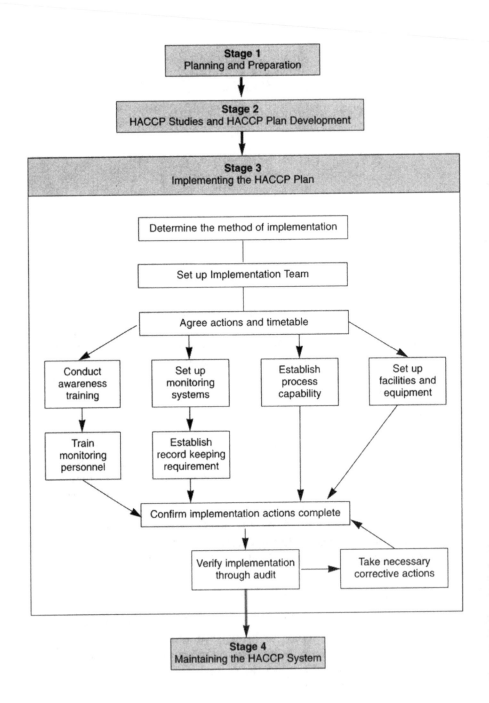

Figure 7.1 Key Stage 3 – implementing the HACCP Plan.

to ensure that all the critical issues can be addressed while working within your constraints. You have already put a large amount of resource into drawing up the HACCP Plan, through the HACCP Team training and making personnel time available for the study. Now it is important to ensure that this resource has not been wasted through improper implementation of the HACCP Plan.

If your processing personnel have limited time available, it is important to ensure that their time is well spent and this usually means being well organized.

There are a number of ways in which a tight budget can be maintained whilst implementing the HACCP System. It is crucial that you do not try to save money by only implementing selected parts of the HACCP Plan. Instead, concentrate on how HACCP can be brought together within your existing operation without additional on-costs. For example, you may not need to create new log sheets for monitoring all the CCPs. It is likely that many of your existing monitoring sheets can either be used directly or amended to take on additional data and/or signature columns (e.g. dating and signing of thermograph charts).

Training is essential for the successful implementation of HACCP, but this does not necessarily have to be done through sole use of external courses. Some of your HACCP Team members could be trained to train other company personnel and conduct briefing sessions. You may also be able to save some time and money by combining training sessions, e.g. food hygiene and HACCP, where the same personnel require different types of training.

If HACCP implementation is carried out to a carefully thought-out plan, it does not need to be a drain on your resources, and instead can help to target resources effectively at those areas which really are critical to the safety of your business. The flow diagram shown in Figure 7.1 is provided as a guide to implementation, and the same project planning techniques as shown in Chapter 3 can be used here. We will follow the main steps in the process through the remainder of this chapter.

The overall approach to implementation may vary but basically there are two main methods – doing it all at once (the 'Big Bang' method), or phased evolution. The former involves waiting until all studies are complete and implementing it in one go, whilst the latter requires that each section of the overall system is implemented as it is completed.

There are advantages and disadvantages of each method (Table 7.1) but the phased method is likely to be the most practical option for most businesses.

Table 7.1 Implementation methods – advantages and disadvantages

Implementation methods 1: the 'Big Bang'

Method	Advantages	Disadvantages

All at once	– Rapid implementation potential
	– Works well in companies with well-established Quality Management Systems
	– Whole work-force involved (more effective than trying to change behaviour of small group within an existing culture)
	– Ease of workforce briefing
		– May take longer period overall than anticipated – all HACCP monitoring and control procedures must be developed before implementation starts
		– No trials of individual system elements
		– Loss of credibility:
		  – If employees see that it's poorly managed
		  – If a CCP fails through lack of support network
		  – Large immediate training requirement
		  – Resource thinly spread, e.g. HACCP Team

Implementation methods 2: phased

Method	Advantages	Disadvantages
Phased	– Quality Management System support elements can be developed as required and alongside – Staged training allows more individual attention – System can be trialled and refined as implementation progresses – More manageable approach → system less likely to fail – HACCP Team resource focused at each stage	– Longer overall implementation timetable – Working with small groups of people in isolation → difficult to change culture – Implementation may lose momentum

7.2 Implementation Team

If you are continuing to carry out HACCP studies then the HACCP Team may not have sufficient time available to devote to implementation. One option may be to set up a separate Implementation Team, the membership of which might include different personnel from the HACCP Study Team. The Human Resources or Personnel

Manager may be very helpful given the training implications and potential changes in working practices. Representation from Production, Engineering and Quality Assurance will also be appropriate and, in a larger organization, you may be able to choose different people from those involved in the initial HACCP studies.

The responsibilities of the Implementation Team will begin with the agreement of the method of implementation and drawing up of an implementation Project Plan (see Chapter 3). The team may also require a budget so finance representation may be helpful. The team should also be responsible for reporting regularly on progress to Senior Management within the company.

Questions to ask when drawing up the implementation Project Plan will include whether you have enough personnel to adequately monitor the CCPs and whether they are the right people for that task. Do your chosen monitors have enough time to fit the monitoring procedures in with their other responsibilities? Have you considered the required level of supervision of CCP monitors, and is this in place? Training will be a key stage in the implementation phase.

The Implementation Team should consider all these issues together with the rest of the management team, in order to identify all the resources required to put together an implementation Project Plan.

7.3 Training

It is often more successful if the education and training of the main workforce is left until the implementation phase. Completion of HACCP studies can take time and, if the workforce is made aware of the HACCP Project at the start, there may be a need for retraining by the time the system is ready for implementation.

Training is fundamentally important if the food safety management system is to control product safety effectively and *all* personnel within a food business should be trained commensurate with their work activities. CCP monitors, their supervisors and managers will need specific training in their role within the HACCP System. Other personnel, including operatives, will need an awareness training as a minimum.

7.3.1 Foundation training

As indicated in Chapter 3, all personnel will need a basic understanding of the HACCP concept, how it will apply to their working environment, and why CCP monitoring is vital. It is also important that all personnel understand the relationship between HACCP and the prerequisite programmes, such as good hygiene practice, where their compliance and support of such routine activities is essential to overall food safety management. If training in hygiene has already taken place, do people need a refresher, and could this now be linked directly to a hazard analysis approach? The Royal Institute of Public Health and Hygiene in the UK has produced an extremely good programme called *The First Certificate in Food Safety* (RIPHH, 1995a) which takes a risk-based approach to food hygiene training for operator-level employees. With specific regard to HACCP, you can probably obtain basic-level visual aids, such as an introductory video, which would facilitate training. Examples of such material are given in the section on 'References, further reading and resource material' towards the end of this book. A practical exercise is also helpful and this can often be delivered by a member of the HACCP Team. As an outline:

1. Using a flip chart, and with the help of the trainees, draw a Process Flow Diagram which represents a simple process. Making a cup of coffee, going shopping, or boiling an egg are all easy options.
2. Again, using a flip chart, transfer the Process Steps to a Hazard Analysis Chart. Get the trainees to brainstorm possible hazards at each step. Depending on their level of prior knowledge, you may need to give a brief reminder of the basic physical, chemical and biological hazards that may occur.
3. For each hazard identified, ask the group for possible control measures.
4. Use the Process Step CCP Decision Tree to show them how to identify which of the control measures are CCPs.

You don't need to do a hazard analysis and identify CCPs through the entire Process Flow Diagram for them to understand the concept, and at this stage you can show them one of the completed HACCP Plans for your business.

7.3.2 CCP monitors

For CCP monitors, their supervisors and managerial staff, you may need to provide additional training.

Monitoring is one of the most important aspects of any HACCP System. This is how we measure that the CCPs are working. CCP monitors therefore play a key role in the production of safe products and they will be able to perform effectively if they understand not only what they are expected to do and why they are doing it, but also how their role fits in with the rest of the HACCP System. An understanding of how essential their role is for the safety of the product is also a key factor in maintaining motivation.

It is vital that all your CCP monitors are instructed in basic HACCP philosophy as previously shown and, in particular, the importance of accurate monitoring. They must understand what the specific hazards are for the CCP in question and how to take corrective action when a deviation occurs. Where appropriate, they will also need to understand the differences between a Target Level and a Critical Limit, and what these values are for each CCP that they are monitoring. In some cases you will need the CCP monitors to adjust the process in order to maintain control and prevent a deviation from occurring. Here it is important to know that your monitor is capable of the required actions. The detail and accuracy requirements for CCP records must also be agreed. In order to achieve this you will have to ensure that training is available for all CCP monitors and this may be carried out by your HACCP or Implementation Team members. Because of the importance of this role, it is recommended that you not only provide training, but also check understanding and competency in the specific task.

New skills may be involved in monitoring, such as taking samples and filling out documentation or keying data into a computer. The involvement of the Human Resources Manager and in some instances the Trade Unions and Works Councils may be necessary where working practices will change as a result of HACCP implementation. CCP monitor understanding and competency may not be gained solely by using a classroom-type training session, it is more likely to be developed through a learning by experience approach, i.e. being shown how to do it and then 'having a go' under supervision.

Both the actual trainer and supervisory/managerial staff will be made much more effective if they themselves have some knowledge of learning styles as they will be able to continue with reinforcement of the training, after the main event has occurred. Their role in this should also be emphasized and made a specific responsibility within their job function. Overall, the training process should be regarded as a motivating experience and shouldn't be conducted in a negative environment. Positive involvement of the CCP monitors is important and this will not usually be gained by dictating rules to them and warning what will happen to them personally if they get it wrong. Obviously any legal obligations are important, but they need to be made aware of their vital role within the food safety management programme as a whole and made to feel part of a team. An additional point to consider is not to forget to train deputy CCP monitors as well, in order to plan for sickness and holiday cover.

7.4 Set up monitoring systems

As a reminder, Monitoring is different from Control and involves conducting tests or observations to confirm that the process remains in control. The monitoring requirements have already been defined within the HACCP Control Chart, here you need to think about the practicalities of implementation within the workplace.

7.4.1 Monitoring records

The monitoring record should have details of the Critical Limits and corrective action procedures. Target Levels within the Critical Limits can be included if the CCP monitor is to adjust the process in order to maintain control. It is also useful to include details of the monitoring method, but this may not be necessary in all cases, particularly when operating within a Quality Management System or where separate work instructions are made available. In drawing up the work instructions it is a good idea if the monitors themselves are asked either to prepare these or to contribute to them. It is better if they are clear, simple documents and the main purpose is usually for ensuring a consistent training approach and for easy reference. The monitoring log sheets themselves should have sufficient space available to record the necessary data and columns should be included for the monitor to sign off and date each monitoring event. In addition, each monitoring sheet should have a cross reference to the HACCP Plan and CCP number. It should be remembered that special CCP log sheets are not an HACCP requirement and existing monitoring sheets might be acceptable. Computer-based records are being used increasingly and can have significant advantages when analysing trends.

An example of a CCP monitoring sheet for one of the CCPs for our ice-cream product is shown in Table 7.2. The 'Reviewed by' section is usually completed by a supervisor or manager.

7.4.2 Methods – use of statistical techniques

A method which may be useful to some organizations when setting up monitoring and verification systems is Statistical Process Control (SPC).

Once a statistical analysis has been carried out for a particular process and has demonstrated that it is capable of achieving an acceptable level of performance, as we saw in Chapter 6, then the statistical profile which has been built up from the capability study can be used to produce a Process Control Chart for the control of a process and its parameters. Such a chart takes the form of data

Table 7.2 Ice-cream – CCP monitoring sheet

Log Sheet CCP No. 11.1	Pasteurizer Automatic Divert Check	HACCP Plan Ref. No. HP001
Monitoring procedure See Work Instruction ID240 – check flow divert operation during cleaning cycle – confirm actions on chart recorder – check sensor against calibrated thermometer	Frequency 2 × daily – at start-up and at shutdown	

Corrective action:
Start-up checks:
– call engineer
– postpone start-up
Shutdown checks:
– contact QA Manager regarding quarantine of product

Date and Time	Result	Action Taken	Signature

Reviewed by:

Title: _____ Signature: _____ Date: _____

capture with graphical plotting of the variations on a time or batch basis. By using the information of the process profile, the process controller (with the aid of a control chart) will be able to tell whether variations in measurements taken of a process parameter are inherent and to be expected as a result of natural random fluctuations of the process (i.e. due to common causes), or whether the variations are of such a magnitude as to be statistically significant and indicate that this shift in the process must be due to some assignable reason (i.e. due to special causes). When a significant variation occurs it indicates that there has been a shift in the overall equilibrium of the process and that an adjustment must be made to restore the process, the shift may also indicate the failure of some plant component. The Process Control Chart is an effective on-line CCP log sheet which is filled in by the operative. The chart gives the operator very rapid notification that the process is going out of control.

Process Control Charts can be used to analyse the process parameters in two respects – mean and range (or standard deviation) – which measure the accuracy and the precision of the process, respectively. The control chart may have upper and lower action limits marked onto them (where appropriate) and can sometimes include intermediate upper and lower warning bands. By taking the mean of the process measurements (say 4 or 5 readings) the operator will get a 'consensus' reading of any overall shift in the process. American Process Control Charts are usually only marked with upper and lower action limits, with no intermediate warning bands. The action and warning limits for the charts are derived from values generated from the process capability analysis and constants extracted from statistical process control tables.

By looking at the range of the individual results used to produce the mean, the operator gets an indication of the stability ('wobble') of the process. Excessive range variation may well indicate the start of plant failure (e.g. a sluggish control valve) – analogous to a spinning top just before falling over. Although the mean reading of all the wobbles may still indicate that it is stationary on its spot, the excessive wobbling (range) would indicate the inherent instability and that the spinning top is just about to fail (i.e. in this case fall over).

The information for a Process Control Chart could be captured on a table as set out in Table 7.3.

A Mean Range Chart would look as set out in Figure 7.2. The basic interpretation of the chart would be that:

- Any result above the upper action level (UAL) or below the lower action level (LAL) should be considered significant and process adjustment should be considered.

Table 7.3 Information gathering for Process Control Charts

Time/Batch No.		08:30	09:00	09:30
Measured values	1	6.5	7.6	8.3
	2	7.6	7.4	7.8
	3	7.5	8.2	7.5
	4	8.1	6.8	7.2
Sum		29.7	30.0	30.8
Average		7.4	7.5	7.7
Range		1.6	1.4	1.1

- Any result between the warning levels (upper and lower) and the corresponding upper and lower action levels should be considered to be suspect; two results in a row in the same band would be considered significant and process adjustment should again be considered.
- Any series of results that show a consistent upward or downward trend should also be considered to be significant.

The use of Process Control Charts is of most benefit when an immediate reading or measurement can be made for assessment in order to achieve instant process control and adjustment.

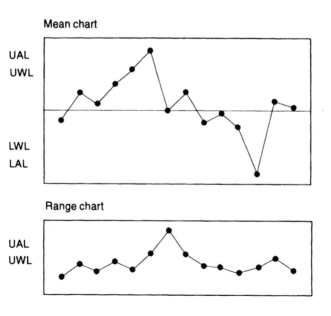

Figure 7.2 A Mean Range Chart. UWL, upper warning level; LWL, lower warning level; UAL, upper action level; LAL, lower action level.

This section has been dealing predominantly with the interpretation of variable data, such as that obtained by measuring time, temperature, flow rate, etc. However, the application of control charts can be used equally effectively with **attribute** data, such as the YES/NO result obtained when checking that a metal detector is working.

7.4.3 Reporting deviations

A deviation occurs when the Critical Limits are exceeded and the CCP goes out of control. The CCP monitors must understand exactly what constitutes a deviation with reference to the Critical Limits. It is essential that they know when to report a problem and who to report to, so the corrective action-reporting structure must be specified. This could be done on the monitoring log sheet or in work instructions.

A deviation also occurs if a monitoring requirement is missed. It can be helpful to measure how well you are doing, not only in terms of operating within either the Critical Limits or Target Levels but also whether all CCP checks are actually being carried out. Some companies choose to report this as two different results and either on a daily, weekly or monthly basis. The method is straightforward and can be a simple percentage figure. For example:

1. Number of CCP checks out of control ÷ number of CCP checks possible × 100 = % out of compliance (use either the Target Level or Critical Limits values).
2. Number of CCP checks missed ÷ number of CCP checks possible × 100 = % checks missed.

7.4.4 Feedback on results and corrective action

The workforce needs to see the whole picture and understand how well the HACCP System is working. This is important from the point of motivation and helps in getting CCP monitors to take responsibility for their part in the proceedings.

Feedback can be given effectively both individually and in groups, and it is always important to stress positive aspects, e.g. telling the monitor that his/her fast action saved the company from financial loss by preventing reject product being produced. This may be done through departmental briefings and performance charts or through written reports being circulated to appropriate personnel. It is also helpful from the motivation point of view if all other staff in the processing area know the importance of the CCP monitor's actions. If the percentage compliance figures are known, this can be presented as a performance chart on a regular basis. This will enable the HACCP Team to better identify trends in the results.

7.5 Record keeping

HACCP Principle 7 requires that effective record-keeping procedures are established to document the HACCP System. Records may be kept of all areas which are critical to product safety, as written evidence that the HACCP Plan is in compliance, i.e. verification that the system has been working correctly. This will also support a defence under litigation proceedings. Records will also be useful in providing a basis for analysis of trends (which in turn may contribute towards improvements in the system) as well as for internal investigation of any food safety incident that may occur. The records do not all have to be in typed format and it is likely that you will hold a number of hand-written documents, e.g. Hazard Analysis Charts and CCP monitoring log sheets. It is also increasingly likely that you will have records in electronic format, in which case they may be more easily archived, but more difficult to prove that they haven't been tampered with in the event of litigation. With paper-based systems it is extremely useful to allocate a unique reference number to each HACCP Plan. This number may then be used on all pieces of documentation relating to the HACCP Plan and cross-referencing of CCP log sheets, monitor training records, etc. will be made easier.

The length of time for which records should be kept will vary, depending on several factors. First, there is likely to be a minimum time for which records must be kept for legal reasons, and this will be determined partly by the country where your operation is located. The record retention time will also depend on the nature of the product itself, e.g. there is little point in keeping records for production of a sandwich with 2 days' shelf-life for as long as the records for production of a canned product which has 4 years' shelf-life. As a general rule, it is wise to keep significant records for at least 1 year following the end of the product shelf-life, although if you have a certified Quality Management System you may need to keep them for a period of 3 years.

The types of HACCP records which might be retained are as follows:

1. The HACCP Plan: as the critical document in the HACCP System the current HACCP Plan should be kept, together with all data collected during its creation. This will mean the Process Flow Diagram and HACCP Control Chart, plus the Hazard Analysis information (Hazard Analysis Chart if used), details of the

HACCP Team which was actually responsible, copies of any CCP Deviations/Non-Compliance Notes and corrective action details. Details of any monitoring procedures need not be retained with the HACCP Plan, providing that they are clearly cross-referenced by number or by location.

2. History of amendments to the HACCP Plan: while it is important to hold the current copy of the HACCP Plan, it is equally useful to have a concise history of any amendments that may have been carried out. Obsolete HACCP Plans will have been destroyed, with the exception of the master copy which should be retained in a secure location.

3. Critical Control Point (CCP) monitoring records: the amount of paper involved in retaining all log sheets may be prohibitive, in which case a monthly/3-monthly summary is recommended. This should clearly detail the CCP number, Critical Limits, indicate any deviations and corrective actions taken, and persons involved. Again, the trend analysis and compliance summaries could prove useful.

4. Hold/Trace/Recall records: in the event of a deviation at a CCP, it may be necessary to hold the product in quarantine pending a decision as to the means of disposition. If the product has been despatched, it will need to be traced and recalled. Records of these activities will need to be retained. It may prove useful in the event of a serious incident if evidence in the form of documented challenge tests on the trace and recall system is available.

5. Training records: evidence that the HACCP Team and other personnel have been trained will almost certainly be needed. A simple record sheet detailing the type and date of training carried out and signed by both trainer and trainee will suffice. Training records should include HACCP training, auditor training, food hygiene training and so on. It is also important that you keep records which demonstrate that your CCP monitors are fully trained and proficient in carrying out their task(s). It is recommended that the Personnel/Human Resources function is responsible for maintaining training records.

6. Audit records: records of HACCP audit through retention of non-compliance notes and reports. This will be discussed in more detail in Chapter 8.

7. Meeting records: it will be useful to keep concise minutes of HACCP meetings. These should indicate any actions required prior to the next meeting, together with the person responsible for taking the action. The minutes will provide a useful focus for the HACCP Team and help to drive the system forward by ensuring that all Team members have the same understanding of what actions were agreed during the meeting.

8. Calibration records: records relating to any instrumentation associated with CCPs. This includes both processing and monitoring
equipment.

9. The HACCP System procedures: you may wish to consider
producing a HACCP Procedural Document for your company as
a way of drawing together all activities associated with the
HACCP programme. It may consist of a HACCP Manual which
contains, first, the company's policy on Food Safety
Management, signed by its most senior executive. The Manual
may then follow on with details of how the company intends to
implement HACCP – the Project Plan – and may contain a master
list of site HACCP Plans and their reference numbers. If you only
have one HACCP Plan, perhaps you may wish to keep the document itself in the manual. A directory of useful external contacts
and training bodies, and perhaps the HACCP audit schedule,
could be considered, along with any master copies of data sheets
and instructions for filling them out.

Regulatory bodies will be interested in reviewing significant
records which relate specifically to establishment of and compliance
with Critical Control Points. However, those companies who wish
to use the HACCP System as a foundation for a quality system will
also need to treat systems-related records as 'significant'.

Records should be stored in an organized manner which enables
easy retrieval.

Figure 7.3 'Retaining records'.

7.6 Facilities and equipment

7.6.1 Facilities

Different facilities are needed for the process itself and for the additional implementation requirements. You should consider the main processing area along with specific facilities required during the process. For example, do you have sufficient handwash basins and are they correctly sited? Will the existing disposal system cope with additional waste from this process? Is there sufficient space for handling the packed product?

You will also need a training facility so that you can brief staff and carry out any specific training required. This may need to be capable of holding large numbers of employees during awareness training sessions, and smaller numbers, for example, during the training of CCP monitors. You may be able to use external training facilities for this purpose.

Additional facilities should be considered, such as a separate test area where log sheets can also be stored, additional computer workstations if electronic data gathering is required, storage for records retention and perhaps a defined location may be needed for work instructions and procedures, e.g. work tables, manual holders on the walls, etc.

7.6.2 Equipment

It is important to establish that you have the correct equipment for each situation. Can it, for example, carry out the process specified and achieve the desired control criteria? Has it been properly calibrated and maintained, and will it be reliable when the HACCP System is implemented? Some important questions to consider at this stage are as follows:

- Do you have the right equipment in place or will you have to buy any new equipment?
- Is it appropriate to the task?
- Is it sensitive enough?
- Can it be calibrated?
- Does it require ancillary equipment, e.g. locked boxes for dud detectors on a can line?
- Is it difficult to operate, e.g. a gas chromatograph?
- Can the CCP monitor interpret the results?
- Will it work on the line or does it require special facilities, i.e. will it withstand the rigours of the production environment?
- Is it cleanable?
- Are there any health and safety constraints?

7.6.3 CCP identification on facilities and equipment

Finally, although not a requirement of HACCP, it may be helpful to identify clearly where the CCPs are physically located within the process. This can be done very simply by the hanging of additional signs within the plant stating 'CCP'. While a very basic element of implementation, seeing signs go up can have a great impact on the workforce and also serves as a constant reminder of which points are CCPs. It is also a visible change which occurs during implementation.

7.7 Implementing the Plan

Having decided on the implementation method ('Big Bang' or phased) at the planning stage, established that the required facilities and equipment are available, trained the workforce and decided on the recording format, the actual implementation of the HACCP Plan simply requires the following actions:

- monitoring CCPs;
- taking action required;
- recording results.

This is where HACCP goes 'live' and managing the CCPs day by day becomes the responsibility of personnel within the operation.

7.8 Confirm and verify that implementation is complete

At the start of this chapter, we looked at the setting up of an Implementation Team. One of the final and important stages in the Implementation Project is to confirm that all actions in the Project Plan have been completed satisfactorily prior to handing the system back to the HACCP Team. In practice, the HACCP Team often becomes a HACCP Maintenance Team and includes strong representation from production personnel.

In Chapter 8, we move on to look at verification that the system is working. A verification audit can and should occur for the first time immediately once the HACCP Plan has been implemented, in order to identify any corrective action requirements as soon as possible.

8

HACCP as a way of life – maintaining your HACCP System

Having completed the HACCP Study and ensured that the CCPs are being monitored, many people breathe a sigh of relief and congratulate themselves that they are using HACCP to manage food safety. But the HACCP Study was completed at a point in time and if it is to remain as effective as it was on the day it was written it **must be maintained** and **verified** as continuing to be effective. Like any Quality Management System, the ongoing maintenance of the HACCP System is where the benefit really lies. The initial study will result in a system that will act as a benchmark for future improvements – driven through identification of weaknesses and by taking corrective action. HACCP should be seen as a way of life throughout the entire company from the moment that the initial studies are completed and the implementation is under way. In this chapter we will consider some of the activities that can drive the system forward, making it live instead of being a set of documents on the QA Manager's office shelf. These include the HACCP Audit, microbiological and chemical testing, analysis of data, awareness of new emerging hazards, ongoing training requirements, and keeping the HACCP Plan up to date. This is the final key stage of HACCP (see Figure 8.1).

The HACCP System must include verification procedures (HACCP Principle 6) to provide assurance that the HACCP Plan has been implemented effectively and that it is being complied with on a day-to-day basis.

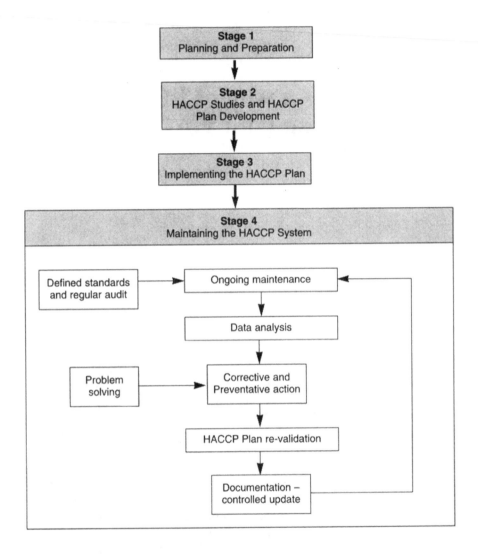

Figure 8.1 Key stage 4 – maintaining the HACCP System.

VERIFICATION:
The application of methods, procedures, tests and other evaluations, in addition to monitoring, to determine compliance with the HACCP Plan.

Some companies may wish to incorporate the verification requirements into the HACCP Plan itself. This can be done by adding an additional column to the HACCP Control Chart or by documenting separately and incorporating into the HACCP Plan file. An example

Table 8.1 Example of a HACCP Control Chart incorporating verification procedures

CCP No.	Process Step	Hazard	Control Measure	Critical Limits	Monitoring Procedure	Monitoring Frequency	Monitoring Responsibility	Corrective Action Procedure	Corrective Action Responsibility	Verification Requirements Action	Verification Requirements Responsibility
	HACCP Module Ingredients:										
1.1	Skimmed milk powder	Antibiotic residues	Effective supplier assurance – Audit	Audit pass	– Audit by trained SQA auditor	Annual	SQA auditor	Change supplier	Purchasing Manager	Send samples to laboratory for antibiotic residues analysis	QA Manager
			Certificate of Analysis – Agreed specification (maximum acceptable levels)	Legal limits	– Check Certificate of Analysis	Each delivery	Incoming goods clerk	Report to QAM Contact supplier Reject consignment	Incoming Goods Clerk QA Manager	Audit the incoming goods, check records monthly	Quality Systems audit team
1.2	Cream	Antibiotic residues	Effective supplier assurance – Audit	Audit pass	– Audit by trained SQA auditor	Annual	SQA auditor	Change supplier	Purchasing Manager	Send samples to laboratory for antibiotic residues analysis	QA Manager
			Certificate of Analysis – Agreed specification (maximum acceptable levels)	Legal limits	– Check Certificate of Analysis	Each delivery	Incoming goods clerk	Report to QAM Contact supplier Reject Consignment	Incoming goods clerk QA Manager	Audit the incoming goods check records monthly	Quality Systems audit team

of the former is given for part of the Iced Delights HACCP Control Chart (Table 8.1).

The verification activities will vary according to the control measures in place.

8.1 Initial and ongoing verification through audit

One of the main methods of verification is through audit. An audit can be regarded as an independent and systematic examination which is carried out in order to determine whether what is actually happening complies with the documented procedures, and also whether the procedures have been implemented such that the stated objectives (safe food) have been achieved. The benefits of auditing a HACCP System will include:

1. Maintaining confidence in the HACCP System through verifying the effectiveness of the controls.
2. Having an independent and objective review of the effectiveness of the HACCP System.
3. Identifying areas for improving and strengthening the system.
4. Providing documented evidence of Due Diligence in managing food safety.
5. Continually reinforcing awareness of food safety management.
6. Removing obsolete control mechanisms.

AUDIT:
A systematic and independent examination to determine whether activities and results comply with the documented procedures; also whether these procedures are implemented effectively and are suitable to achieve the objectives.

In HACCP terms, achieving the objectives means managing the manufacture and distribution of safe food products through use of HACCP.

The audit can be considered as a 'health' check of the HACCP System. It is a means of determining its strengths and weaknesses and, by taking appropriate corrective actions, a route to continuous improvement.

8.1.1 Types of audit used in HACCP

There are three main approaches to auditing a HACCP System.

(a) The Systems Audit

If you have chosen to manage HACCP using a Quality Management Systems approach, that is, against each of the HACCP Principles, defined procedures are in place which state precisely how HACCP will be implemented and maintained, the systems audit may be used. The purpose of the audit is to find any weakness in the system and to ensure that corrective action is taken. This will entail taking a thorough, systematic and independent review of all or part of the HACCP System. Priorities for corrective actions can be assigned against food safety risk. For example, if you have a clearly defined requirement for a HACCP Team approach, the auditor may want to look at the team structure, team member qualifications and training records, details of who had carried out the HACCP Studies – one team member or with full team input. Both current and historical documentation will be reviewed. This type of audit is most commonly used for ISO 9000 series (Quality Management System) audits.

(b) The Compliance Audit

Again, the audit will be independent, but usually involves a more focused, in-depth inspection of the operation against the standards defined in the HACCP Plan. This type of audit will be most commonly used for HACCP, from checking CCP compliance to ensuring that the HACCP Team had originally identified the hazards correctly along with the appropriate controls in the process. In the latter case, the compliance audit will be done either by an internal or external HACCP audit expert.

In summary, the HACCP Compliance Audit could be assessing two areas:

- compliance with the requirements of the HACCP Principles;
- compliance with the documented HACCP Plan – has it been implemented properly and is it still correct?

(c) The Investigative Audit

This is an investigation into a specific problem area. This type of audit may be used when a CCP regularly goes out of control – investigating the real cause in order to take corrective action, or where a previously unknown problem has arisen.

In implementing and maintaining HACCP, all three types of audit may be used, either on their own, or in combination. Whatever type of audit is used, the essential elements will remain the same.

8.1.2 Identification and training of auditors

HACCP auditors must be skilled in the techniques of auditing, knowledgeable in HACCP itself and technically qualified in the area under study. For this reason it is often advisable to use members of the HACCP Team as auditors, as many of the required competencies will be the same. However, it is also important to have a degree of independence and therefore it can be an advantage to use someone who wasn't on the original HACCP Team and/or a representative from another discipline. They may be more inclined to challenge existing practices and beliefs than HACCP Team members who are closely involved with the system.

For 'in-house' audits, care must also be taken to ensure that the auditors do not audit their own departments. You could use external specialists such as HACCP experts or, alternatively, could work together with your customer technologists or regulatory authorities, if this is appropriate. In larger manufacturing sites with several HACCP Teams it can be helpful to have them audit each other's HACCP Plans.

Audit techniques can be taught fairly quickly through attending an auditor training course and by shadowing experienced auditors. If the auditor is inexperienced in hazard analysis and HACCP techniques, then the training period will take considerably longer. Where an audit is to be conducted by more than one person, the responsibility for leading the audit should be defined.

8.1.3 Scheduling of audits

It is essential that an audit schedule is established. You will want to ensure that the scope of each audit is clearly defined in order that the entire HACCP System is reviewed and no element missed out. It is recommended that, following an initial verification that the HACCP Plan has been implemented, a 3-monthly audit of the CCPs would be reasonable. It would be possible to perhaps schedule audits of part of the system on a weekly or monthly basis. The frequency will depend on the nature of the business, for example a seasonal vegetable packer may only audit once a year whereas a ready-meal factory with frequent menu changes may audit monthly. The schedule needs to be established so that auditors can be assigned to it well in advance.

Let's now consider the steps that will be required in a HACCP Compliance Audit (Figure 8.2). This audit guidance is based on doing a first time, third-party audit, e.g. of a key (CCP Sensitive Ingredient) supplier, or a HACCP Plan verification audit for a different department within a larger company, but it will also be useful for those wanting to conduct internal audits as verification of their own systems. You should adapt this for your own use depending on circumstances.

We will now take each of these stages and look in detail at what happens.

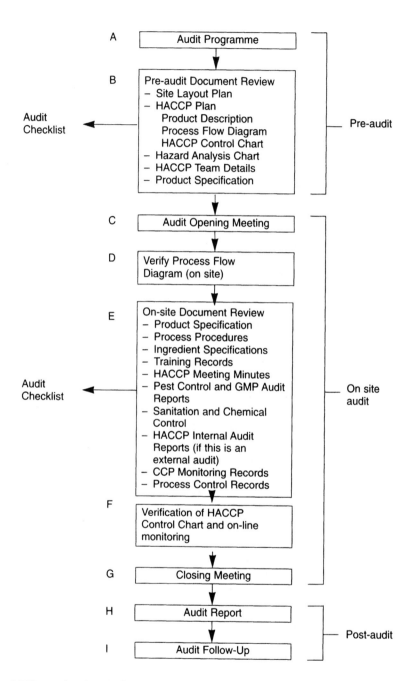

Figure 8.2 Example of typical steps in a HACCP Audit.

A. Audit programme

It is useful to prepare an agenda for the audit programme. This will serve to notify personnel who may be required during the audit of your intended timetable and for them to ensure that they are available. Include the start and finish times.

You will need also to make sure that you have all documentation required for the pre-audit review. In alerting the auditees to the agenda for the audit, you will be able to request all relevant documentation, as indicated.

B. Pre-audit Document Review

Before the audit, all documentation relating to the scope of the audit can be reviewed by the auditors. This will be a very important part of the audit as an initial audit checklist can be drawn up during this process.

What documentation should be reviewed? In answering that question, let us consider what would be available. Firstly, the site layout plan, which is useful for understanding both the flow of product through the site and also the scale of the operation, including other products produced. Secondly, the HACCP Plan. Start with the Process Flow Diagram and product specifications relating to it. Compare one against the other, noting whether all elements of the corresponding specifications are included in the Process Flow Diagram and vice versa. Consider whether all time/temperature information is adequately covered by the Process Flow Diagram.

The pre-audit Document Review can be done as an initial scan – to get a feel for who carried out the HACCP Study, the style, completeness, and also familiarization with the site being audited, and the product and process itself. It will also give you an opportunity to carry out some research before the audit.

If you are auditing a HACCP Plan for the first time (perhaps as part of a verification exercise), an important part of your audit will be to assess the competency of the people responsible for the study. One way of doing this is to take sections of the Process Flow Diagram, preferably a high-risk section, and without reference to the HACCP Control Chart carry out your own hazard analysis based on your expert knowledge and use of reference material, legislation, etc. Having done that, compare your result with the original Hazard Analysis Chart. At this stage, too, you will be able to consider whether food safety hazards only have been included or whether quality and legal hazards have been identified. Also, whether the hazards and control measures are defined precisely or are rather vague and general – is there a control measure for each specific hazard? Has the hazard analysis been carried out in an organized manner or is it a jumble of hazards and control measures? Following on from this, you will begin to be able to judge how the CCPs have been established. Make a note to check whether records were kept of the decision-making process.

You may decide to assess the competency of the HACCP Team by then taking the CCP decision trees and using your own expert knowledge, determining where you think the CCPs are, and why.

In looking at the HACCP Control Chart, make sure that food safety hazards are clearly and separately identified. Considering each of the columns on the chart, is the corrective action identified going to be effective and is it realistic? What about the people responsible for monitoring and taking corrective action? Make notes of who they are so that you can talk to them during the audit. For example, if the 'Goods-Inwards' clerk is defined as being responsible for checking Certificates of Conformance for certain high-risk raw materials, you will be able to question him or her to assess whether he/she has been trained, what his/her terms of reference are, where his/her CCP log sheets are kept, who reviews them and so on. The same approach can be taken for each CCP. Consider the monitoring procedures and frequency – perhaps a Certificate of Analysis of a *Salmonella* test of an ingredient is specified and the procedure for checking this is cross-referenced by a reference number. You will be able, during the audit, to check that the procedure exists, and that the issuing laboratory has been validated.

You should by now have a great many questions that you will want to ask during the audit. You may also want to discover what steps were taken to capture any information relating to process control points. This is a useful indicator of the approach of the HACCP Team – are they using HACCP to effect business improvement in addition to product safety?

If you feel that the Document Review has indicated obvious inadequacies, it may be advisable to stop the audit at this point. The deficiencies should be discussed with the HACCP Team, who can then review their HACCP System and implement any further training requirements.

Audit checklists
One of the most important aspects of an audit is the required organized approach to its execution. Many people find that using a checklist is helpful during an audit and the Process Flow Diagram itself might be useful in drawing this up. One possible audit checklist format is shown in Table 8.2. The 'Considerations' column can be completed during the Document Review for each step of the process, and the 'Auditors' findings' column during the audit itself.

Table 8.2 Example of audit checklist

Process step	Considerations, questions and points to raise on site	Auditor's findings
Raw materials (including goods inward and storage) (List raw materials here) →		
Process (List process steps here) →		
Packaging and despatch (List packaging steps here) →		

As an example of the type of operations the auditor might add to the checklist, the following (non-exhaustive) list is provided:

Raw materials

- Are the 'critical' ingredients identified?
- Are they being handled according to specification?
- Storage conditions – are they as stated?
- Are all raw materials and process/storage activities included in the flow diagram? (Rework can be included as an ingredient.)
- Are positive release or quarantine requirements being adhered to?
- Are agreed specifications in place?

Supplier Quality Assurance

- Have auditors been trained, and how?
- Have all suppliers been audited?
- Have suppliers changed since HACCP Plan development?
- Check visit reports – has all corrective action been followed up as required?

Certificates of Analysis and Conformance

- Are these being used?
- Do goods-receiving operators know what to do with them?
- Have they been checked as accurate?

Process

- Have all activities been included?
- Is the Process Flow Diagram correct?
- Have process capability studies been carried out?
- Have any changes been made since the Process Flow Diagram was drawn up? If so, how does the HACCP Team get notified and how were the changes recorded and approved?
- Were any changes discussed with the HACCP Team before implementation?
- Are there rework opportunities and have they been included?
- What methods were used to ensure the accuracy of the Hazard Analysis?
- How were Critical Limits established?
- Are the HACCP records clearly identified by unique reference numbers?
- Are all documents accurate and current?
- Is monitoring equipment calibrated?
- Have CCP monitors been trained?
- Are CCP records being reviewed? By whom?
- Is the information on the HACCP Control Chart accurate?

- Are time/temperature parameters being achieved?
- Are CCP log sheets being filled out correctly?
- Is frequency of monitoring adequate?
- Has corrective action been recorded and has the effectiveness been verified?
- Have statistically valid sampling plans been drawn up?
- What is the general standard of GMP and other prerequisite programmes such as Pest Control, Chemical Control and Allergen Control?
- Is there a hygiene schedule?
- Are production codes legible on the packaging?
- Are customer usage instructions clear and accurate?
- How was shelf-life determined?
- Are the packaging materials as specified?
- Are Statistical Process Control records being used to demonstrate that the process is in control on a day-to-day basis?
- Do records agree with stated activities?
- Are there any cross-contamination opportunities?

Packaging and despatch

- Are the packaging materials as specified?
- Are storage conditions as stated?
- Are distribution procedures in-house or third party?
- Are good distribution practices being maintained? Check hygiene, handling and temperature if chilled or frozen.

This list is not exhaustive but, as with hazard analysis, there are many areas to be covered.

The above activities can happen in advance of the audit itself. Let's now consider what happens on the audit day(s).

C. Opening meeting

It is good audit practice to begin with a brief opening meeting. Use it to confirm with the key people being audited (auditees), the audit scope, timetable and personnel required. Confirm too, the time and location of the closing meeting and who will be needed. Request any additional documentation required for the on-site Document Review.

D. Verify Process Flow Diagram

It is essential to verify the Process Flow Diagram at an early stage, unlike other Quality System audits, where other aspects of the documentation may be considered before going into the factory or process operation. This is simply a matter of walking through the process from start to finish. However, it may take some time and should not be hurried. Stand and observe what is happening in each area.

The auditor's tools of eyes, ears and mouth are essential to:

- **watch** what is going on;
- **listen** actively to what people are saying;
- **ask** questions, talk to operatives; for example, ask them what they are doing. Do they always do it that way? When might they do it differently?

Check for evidence of any time/temperature stages. Look for opportunities for cross-contamination. What about holding periods? Could there be time enough for toxin formation or spore germination? This may be particularly relevant to high-risk raw materials or part-made product where there is a high degree of handling. Make a note of people with whom you have spoken, check their training records during the Document Review.

You may want to pick up a few finished product codes out in the stores during your Process Flow Diagram verification and use this to trace test results and records, and to test traceability procedures during the on-site Document Review and factory audit.

E/F. On-site Document review and verification of the HACCP Control Chart

Having established the Process Flow Diagram status, you will be able to carry out a thorough on-site Document Review from a more informed base. As with the Process Flow Diagram verification, use eyes, ears and mouth to search for evidence of compliance with the HACCP Plan. This time you should include a full review of operational procedures for CCP monitoring, CCP monitoring records, training records, etc. Check the prerequisite GMP and hygiene maintenance records, pest control and also the HACCP Team meeting minutes. In the latter case, it may be helpful to get an idea of the decision-making process, who attended the meetings on each occasion, how often they occurred and whether difficulties were encountered.

The review will also include previous internal audit records where non-compliances may have been found. The assurance of the effectiveness of any corrective actions taken must be sought. Other quality- and safety-related data for review will include a review of customer complaints, customer audit reports, and any minutes of HACCP or Quality Improvement Project meetings relating to the audit. The on-site Document Review is aimed specifically at verifying that the HACCP System is working effectively in the workplace.

At this point, the audit process is in the conclusion stage. A few key points to note:

- Final investigation of any anomalies found during the audit process.
- Note any points of concern that cannot be resolved.
- Ensure that identified deficiencies are clearly understood and

Table 8.3 Example of non-compliance note

HACCP Audit Non-compliance Note	No:
Location:	Date:
Area under review:	HACCP Plan Ref. No.
Non-compliance:	
Action required by (date):	
Auditors: 1. 2. 3.	
Accepted by Auditee:	
Corrective action:	
Verified (Auditor) Date:	

supported by evidence (specific examples of corrective action not being followed up, for example).
• Communicate any deficiencies at the time of discovery and obtain agreement.

G. Closing meeting

This is the first opportunity to present the audit findings and give an overall view of the proceedings. Non-compliances should be discussed together with supporting evidence and a schedule for major corrective actions agreed. The recommended corrective actions should be generated by the auditee and agreed by the Departmental Manager. It is important that recommendations are feasible. An example of a non-compliance note is shown in Table 8.3. This type of record can be used to document the outcome of the audit.

H. Audit reporting

Audit reports should provide evidence of the findings of the audit – primarily what deficiencies have been found in the HACCP System.

While non-compliance notes should be issued ideally on the day of the audit, it may be appropriate for the auditor(s) to issue an audit summary report. This may be useful to company management and also to the HACCP Team and subsequent auditors.

Again, a pro forma might be a useful means of summarizing. An example is given in Table 8.4. The 'Additional comments' section can be used to note any observations that may not have resulted in a non-compliance note but where minor corrective actions are perhaps needed.

I. Audit follow-up

Outstanding non-compliance notes may be discussed at HACCP Team meetings and, if seriously impacting on food safety management, by senior management or board meetings in order to ensure that timely corrective action is taken. Non-compliance notes should be closed and signed off as soon as the corrective action has been taken. Even so, they will need to be reviewed during any subsequent audit to ensure that the corrective actions taken have been effective on an ongoing basis.

8.2 Analysis of data

The HACCP procedures will generate a number of records which should be reviewed on a regular basis as part of verification. There is nothing worse than seeing records of useful measurements, i.e. data,

Table 8.4 Example of a HACCP Audit Summary pro forma

HACCP Audit Summary	
Location:	Date of Audit:
Audit Ref. No:	Area under review:
Auditors:	
NCN Ref. No.	Summary of Non-compliances
Additional comments:	
Signed Auditor(s)	Date:
Circulated to:	

pile up in the QA Manager's office, not being used to make process improvements and with no analysis carried out. Some suggested answers to common questions on this data analysis are as follows.

Why analyse data?

- To verify that the HACCP Plan continues to be effective.
- To enable trends to be recognized and corrective action teams to

be set up to deal with the cause, e.g. customer complaints and recurring CCP deviations.

- To launch investigatory audits of problem areas.
- To ensure that timely corrective actions are being taken through trace audits of meeting minutes.
- To demonstrate that the supporting prerequisite programmes of GMP, hygiene/cleaning, pest control, etc. are in control.

What data should be available?

- CCP log sheets.
- Finished product test results
  - microbiological
  - chemical
  - physical.
- Process Control Charts.
- Audit reports
  - non-compliance notes
  - corrective action reports.
- Minutes of food safety-related meetings
  - HACCP Teams
  - hygiene
  - quality review.
- Pest control records.
- Consumer and customer complaint data.

How often should data be reviewed?
Here are some suggestions:

- Daily
  - CCP log sheets
  - Process Control Charts.
- Weekly
  - finished-product test results
  - environmental microbiological test results.
- Monthly
  - customer complaints reports
  - hygiene meetings.
- Three-monthly
  - CCP deviation summaries
  - corrective action reports
  - audit reports

- HACCP and quality meetings
- pest control records.

- Annual
  - audit reports
  - minutes of food safety meetings
  - customer complaints trend.

Who should review it?

The following, as appropriate:

- HACCP Team
- Quality Manager
- Operations Manager
- Operations Supervisor
- R & D Personnel
- CCP monitoring records should be signed off both by the CCP Monitor and by a responsible reviewing official. This is a requirement of Codex (1997).

How should the data be analysed?

It is important that the information available is used to provide verification that the HACCP System and its support network or pre-requisite programmes are working effectively. The actual analysis of the information will be made much easier if handled electronically, though all businesses may not be able to do this. Whichever method you choose, the analysis should clearly indicate trends. Use of graphs and charts can provide an easily interpreted visual record that can be shared with the workforce. Where performance indicators are used, these can be plotted on a graph

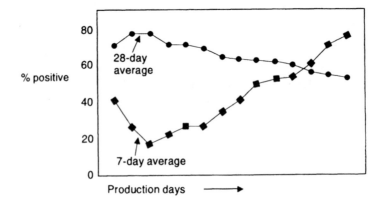

Figure 8.3 Moving average control chart.

to indicate performance, likewise with complaints data. The use of statistical techniques, as seen earlier (Chapter 6), can also be utilized.

For the identification of microbiological trends, on a more retrospective basis and as part of the verification procedures, the use of a rolling (moving) average of percentage of samples that were either present/absent per unit weight, or the percentage of samples with counts that were either less than or greater than a specification per gram, can be particularly effective at picking up trends and eliminating fluctuations (Figure 8.3). It is however also important to respond to and investigate each individual incidence of unacceptably high microbiological results. These rolling averages can be calculated on a weekly, monthly or quarterly basis (as found to be most appropriate) and associated with each rolling average can be an assigned warning level. This may be particularly useful for monitoring the effectiveness of a cleaning schedule.

The use of the principles of Statistical Process Control can be a very powerful tool in the implementation of HACCP, for ensuring that the Critical Control Points are being effectively monitored and controlled on an ongoing basis and for the evaluation of critical hygiene data (Hayes *et al.*, 1997).

8.3 Keeping abreast of emerging hazards

Having established your HACCP System, you will need to ensure that you are kept up to date on new emerging hazards which could have an impact on your product and require modification of the HACCP Plan. Why will new hazards arise? Let us consider just a few of the possible answers to this question.

1. **New technologies.** This could cover a wide range of activities, but a few recent (within the past 20 years) areas to consider are irradiation, microwaving, 'sous vide', mycoprotein and advances in aseptic packaging, modified atmosphere packaging (MAP) and extrusion technology. Each brings its own hazards and risks.
2. **More natural foods.** Consider the ongoing trend particularly for fewer preservatives and more natural ingredients.
3. **New combinations of foods.** For example, chilled ready-to-eat sandwiches containing unusual combinations of fish, fruit, meat, nuts, eggs, vegetables, mayonnaise and conserves, etc., where the interface between foods may present an opportunity for microbiological growth that was not there in the individual components.
4. **Changing legislation.** The banning of additives such as ethylene oxide for the treatment of herbs and spices may have been

beneficial from a chemical safety viewpoint but presented difficulties as new methods had to be developed to reduce microbial loading.

5. **New information on existing issues**. Keep updated on information regarding, for example, bovine spongiform encephalopathy (BSE) in cattle, causes of microbiological food safety incidents, increased understanding of micro-organisms and methods for their detection and increasing importance of emerging pathogens. Also, the results of any government surveys and research programmes that may be relevant.

6. **New ways of presenting food to the consumer**. Many examples could be considered here, as companies are always looking for ideas that will increase market share through changes in product usage or to meet a new demand which has arisen through our changing lifestyles. Consider the enormous growth of chilled ready meals, new sales outlets such as garage forecourts complete with microwaves, the restaurant trend for warm meat or fish salads, the trend for shopping on a weekly basis instead of daily.

Information on such matters can be obtained through Food Research Associations, many of whom regularly circulate abstracts of newly published information. Access to a good reference library may also be helpful. Otherwise use your customers and suppliers as a source of information which is likely to be highly relevant to your products and market. Industry symposia can also be a useful way to meet people with a similar interest. In addition, experienced consultants and data published by the government and media can be used. Increasingly, the Internet is also proving to be a good source of up-to-date information (see section on References, further reading and resource material).

8.4 Updating and amending your HACCP Plan

The HACCP Plan will need to be updated and amended periodically to ensure that it remains current. This is only really common sense – a HACCP Plan which was drawn up a year ago is unlikely to reflect current activities accurately. In the real world, manufacturing operations change and do so for various reasons; for example, new raw material types, changes to process controls, CCP deviations, new methods to improve production efficiency, new equipment, factory structural changes, new packaging, extended shelf-life, new emerging hazards and so on. The HACCP Audit may also provide reasons for change but remember that the audit is only a sampling exercise, an indicator of whether the HACCP Plan is being complied with and is correct. In addition to this, there may be new information that would require a review of the critical limits, for example new data on toxicity of chemical hazards, infectious dosage of microbiological hazards, or legislative changes.

It is recommended that periodic revalidations of the Plan are also carried out. Revalidation can be considered as a complete review of the HACCP Plan in order to confirm its accuracy. This is usually done by the HACCP Team and should be done at least annually. An in-depth inspection is carried out of all HACCP Control Limits and documentation. It is a good opportunity for the HACCP Team to consider the effectiveness of the HACCP System, and to determine what new approaches may be needed in the year ahead.

Clearly, when all HACCP Plans have been implemented the HACCP Team will still be required to meet in order to discuss maintenance, but this will be on a less frequent basis. It is important that HACCP Team members do find the time to participate in maintenance meetings, and it is therefore best to keep meetings as short as possible. Many companies find that meetings of duration between half an hour and an hour work best.

8.4.1 Controlled amendment

Any changes made to the HACCP Plan will need to be recorded and approved. The revalidation exercise should also be recorded even if no changes to the plan were needed. This may provide useful due diligence evidence in the event of a prosecution or customer audit.

One of the reasons for failure of HACCP to control safe food production is because the HACCP Plan is a paper (or electronic) documented system and although the HACCP Team itself may be meeting regularly, and diligently amending it when new hazard information appears, they may not be receiving notification of changes to the product formulation or process, which in turn will not result in an update of the HACCP System. How can this situation be avoided? Product formulation and ingredient changes can be controlled through the design function, and use of a more structured approach such as the product safety assessment discussed in Chapter 5 can ensure that the changes are captured and analysed for possible hazards, or to see whether they eliminate any existing ones. Process changes such as equipment modification can be more difficult to capture, and this really requires the complete support of engineering, production and other key functions within the business. Use of a 'HACCP Change Request' pro forma can sometimes be helpful. An example is given in Table 8.5.

Table 8.5 HACCP Change Request form

HACCP Change Request Form	HACCP Plan Ref. No.
Details of change:	
Submitted by:	Date:
HACCP Team Assessment:	
Action required:	
Authorised by:	Date:
HACCP Team Leader	
Copied to: HACCP Team Leader Originator	

The HACCP Team Leader on receiving the HACCP Change Request form can discuss the likely issues with the team. Again, they will conduct a hazard analysis to determine whether any new hazards arise or whether any existing hazards can be eliminated.

Using the Iced Delights example, we will now look at the activities that occurred when the company decided to introduce a new flavour variant – chocolate peanut ice-cream. Their review process can be divided into three categories: product safety assessment, HACCP Plan amendment and prerequisite review.

Table 8.6 Product safety assessment – chocolate peanut ice-cream

PRODUCT Chocolate peanut ice-cream			FORMULA CP1				DATE 5-11-97	
Stage	Considerations	Criteria	Likely hazard	Control measures	Is control possible?	Validation of control	Recommendations to HACCP Team
Concept	Targeted at general population including high-risk groups.	Frozen product to be eaten without any further process	Vegetative pathogens with low infective dose **Allergic reaction from vulnerable groups**	Pasteurization, filtration, supplier assurance **Labelling**	Yes **Yes**		**Check labelling for presence of the nut ingredient**
Ingredients	Sensitive ingredients and supplier control						
	SMP	Dried	Antibiotic residues	SQA	Yes	Supplier's antibiotic monitoring procedures	Verify that antibiotic monitoring procedures satisfactorily covered during SQA audits
	Cream	Pasteurized, chilled		SQA	Yes		
	Chocolate chips	Ready to use	*Salmonella* Pesticide residues	SQA	Yes	Certificates of Analysis received with each batch	Careful control at supplier → effective supplier management (including audit of processing site and microbiological test facilities). This is a CCP

Table 8.6 *continued*

PRODUCT Chocolate peanut ice-cream			FORMULA CP1			DATE 5-11-97		Page 2 of 4
Stage	Considerations	Criteria	Likely hazard	Control measures	Is control possible?	Validation of control		Recommendations to HACCP Team
	Water	Mains	Protozoa	Supplier control	Unknown	Legal obligation		Ensure proactive relationship with water authority
	Stabilizer	White powder	No hazard identified	Labelling in plant	–	–		Controlled labelling of white powder ingredients must be in place in the factory
	Packaging	Plastic tubs and film	Plasticizers and additives	SQA	Yes	Supplier testing results		Ensure product suitability testing has occurred and is documented as complying with legal requirements
	Peanuts	**Chopped and shelled**	**Aflatoxins Nutshells**	**Supplier control Supplier control**	**Yes Yes**	**Certificates of Analysis with every batch**		**An audit of the supplier will be necessary**
Legal	Ingredients/ product	Thermal process control Recipe	Food safety		Yes	Regulations as per manufacturing country		Check compliance

Stage	Considerations	Criteria	Likely hazard	Control measures	Is control possible?	Validation of control	Recommendations to HACCP Team
Recipe/intrinsic factors	a_w, pH, chemical preservatives, organic acids – none will control product safety	Insufficient sugar to prevent micro growth totally	No – product is frozen	–	–	–	–
Process	Process conditions	Pasteurization failure	Survival of vegetative pathogens	Correct heat process	Yes	Required	Ensure that the effectiveness of the heat process is validated for this formulation. Critical limits will need to be established
		Temperature control	Spore outgrowth	Effective temperature control and stock rotation	Yes	Audited on a monthly schedule. Calibrated temperature recording already in place	None
	Contamination	Air filtration failure	Introduction of pathogens	Effective filtration	Yes	Required	Check filter size and performance criteria. Microbiologically filtered air necessary

Table 8.6 continued

PRODUCT Chocolate Peanut ice-cream			FORMULA CP1				DATE 5-11-97	
Stage	Considerations	Criteria	Likely hazard	Control measures	Is control possible?	Validation of control		Recommendations to HACCP Team
	Contamination	Nut contamination of processing equipment. Incorrect label	Allergen contamination of other products where labelling would not provide controls. Nuts not properly identified	Additional cleaning and dedicated equipment where possible. Check label	Unknown / Yes	Residue testing of rinse waters and of first following product through the line / Audit labels		No production trials can take place until this has been validated as other products could become contaminated / Label check will be a CCP
Post factory	Shelf-life	Product consumed beyond shelf-life	No hazard identified	–	–	–		–
	Customer abuse	Temperature abuse	Unlikely – sufficient abuse for growth will render product inedible	–	–	–		–
		Contamination with serving spoon	Unlikely – only low numbers; will not grow in freezer	–	–	–		–
			Slight risk perhaps from leaving serving spoons in water between servings	None possible	No	–		The product is targeted at the domestic market rather than catering, therefore hazards associated with mass servings are unlikely to be realized. Revisit if a 'catering' version is launched

Signed: J. Smith

(Position) Development Manager

Date: 5-11-97

(a) Product safety assessment

Here the Iced Delights HACCP Team identified additional hazards for the new product itself (aflatoxin and nutshell, allergic reaction if wrongly packaged/labelled) and all other existing products in the range (allergic reaction through cross-contamination). Table 8.6 shows how this information was captured. Bold type indicates the additions made to the table since the previous assessment.

(b) HACCP Plan amendment

The Iced Delights HACCP Team used the information on the product safety assessment to go on to complete the amendment of the modular HACCP Plan. First, they reviewed the Process Flow Diagrams and then a Hazard Analysis was undertaken starting with Module HM8, Dry Particulate Preparation. They saw that the subsequent module (which Module HM8 linked into) would need review also, namely the Filling Room HM12. The HACCP Team needed to assess physically, through on-line audit, where cross-contamination was likely to occur and whether additional control measures were needed in order to support the existing cleaning schedule, which was also reviewed for effectiveness. During this stage, validation trials were conducted and cleaning rinse waters sent away for external analysis. Additional filling heads were found to be necessary due to the difficulties in cleaning them, which eventually meant that an additional CCP had been identified. Production scheduling was also felt to be necessary, such that the peanut product was always scheduled last, before a major clean down, and the product following down the line from the peanut variety would be tested for peanut residue. The section of the amended HACCP Control Chart is shown in Table 8.7.

Prerequisite programme review

We have already seen that the cleaning schedule was reviewed by the HACCP Team. This would have been strengthened to include additional verification controls on the general cleaning. Also, observations of physical cross-contamination (i.e. nut spillage) would need procedures and training on how to clean these up. These additional controls will be fully assessed in order to see whether any are regarded as critical, i.e. CCPs. Other prerequisite reviews will include occupational health procedures where existing and future employees will need to be screened for their own susceptibility to peanut allergic reaction.

Table 8.7 Amended HACCP Control Chart for ice-cream with addition of peanut hazard controls

HACCP Plan Ref. HP001/2				Iced Delights HACCP CONTROL CHART					Date: 14-11-97 Approved by *A. Jones* HACCP Team Leader				Supersedes: 5-11-97	
CCP No.	Process Step	Hazard	Control Measure	Critical Limits	Monitoring				Corrective Action		Verification Requirements				
					Procedure	Frequency	Responsibility		Procedure	Responsibility	Action	Responsibility		
	HACCP Module 12: Filling room													
12.1	Date coding	Inability to trace and recall product resulting in unfit product in marketplace	Effective date and batch coding	Correct code applied	Visual inspection	Start-up and half-hourly	Production Operator		Quarantine product and recode Input correct code or repair as appropriate	Production Supervisor/ Operator	End of day review of records	Line Manager		
12.2	Metal detecting	Metal in packed product	Effective metal detection	2.0 mm ferrous, 3.0 mm non-ferrous, 4.0 mm stainless steel	Check metal detector with test pieces	Start-up and half-hourly to include end of run	Production Operator		– Repair/ recalibrate metal detector – Quarantine and recheck product back to previous good check	Production Operator QA Supervisor Engineer	4-hourly audit of records	Quality Assurance technician		

No.	Process step	Hazard	Control measure	Critical limit	Monitoring method	Frequency	Responsibility	Corrective action	Responsibility	Verification	Responsibility
12.3	Filling	Allergen cross-contamination	Change to unique filling head (Peanut 1)	Must be in place for the peanut variety only	Confirm presence Confirm removal	At start-up of peanut variety At change-over to other varieties	Maintenance Supervisor Maintenance Supervisor	Stop line and change to correct filling head Stop line, put any production on hold	Engineer Production Operator	Rinse-water testing of the filling operation for peanut residues at changeover CIP Residue testing of next product through the line	Quality Assurance Manager
12.4	Labelling	Absence of nut identification *nb	Confirm correct labelling	Present	Visual inspection	Start-up and half-hourly to include end of run	Production operator	Await correct labels	Production operator	End of day review of records	Line Managr
								Hold product to last good check	QA supervisor	"	Line manager

*nb To ensure labels are consistently correct throughout the run, bar code scanners may be introduced as an additional control measure either at the supplier premises or at Iced Delights.

Table 8.8 History of Amendments sheet

| HACCP Plan Reference: .... |
| Page: .... |

Date	Amendment	Reason	Approval signature

Once the HACCP Team has completed its assessment and the HACCP Control Chart updated, all related procedures, such as those used for monitoring and verification, will need also to be updated to include the change. This is where use of a Quality Management System such as ISO 9001 can be really helpful in providing the framework for control of documentation.

8.4.2 Capturing amendments to the HACCP Plan

A useful method of recording these activities is to draw up a History of Amendments sheet. This may be the reverse or a second page to the HACCP Plan approval sheet, if this is a separate document. The main elements to include are shown in Table 8.8.

8.5 Ongoing training requirements

In Chapter 3 we considered initial training needs, and in Chapter 7 the implementation training needs, but what about ongoing training requirements?

8.5.1 Refresher training

It is important that the company updates and refreshes its approach to HACCP on an ongoing basis. A year or two on from implementing HACCP, the company will almost certainly have begun to develop its own interpretation of the HACCP Principles. It will be useful to keep up to date with current international thinking through attendance at industry seminars and literature surveys. This obviously does not need to be done by everyone in the company; most likely it will be a HACCP Team member who, on returning, should use the new information to brief other company employees.

CCP monitors will also need refresher training in order to maintain their understanding of the HACCP System. This too can be done through internal briefings or by posting information on noticeboards.

8.5.2 Training new HACCP Team members and CCP monitors

This task will get easier as the company becomes more familiar with HACCP. Personnel changes will make it necessary for new people to come into the HACCP Team or new CCP monitors to be appointed. These new people will not have the advantage of being involved from the beginning so care must be taken to ensure that they have the same level of understanding as their colleagues. HACCP is a team activity and it may be useful when appointing new HACCP Team members to go back through a team-building exercise. This will help to establish the new team relationship – trust and interdependence will not automatically appear with the new member.

8.5.3 Training new staff

In addition to new HACCP Team members and CCP monitors, the company staff turnover must be considered. At all levels and disciplines of staff, HACCP training will need to be carried out appropriately, from spending a whole day briefing a new board director to an hour of awareness training with a cleaner.

8.5.4 HACCP Plan Amendments training

CCP monitors, and their supervisors, may need to be trained following a change to the HACCP Plan. It is important that they are aware of what the change is, why it occurred and what it means to their activities.

This will also help to encourage them to give feedback regarding any areas that are not quite correct as they will see that the HACCP System is a 'live' system which relies on input from all personnel in order to make it work effectively.

8.5.5 Ongoing awareness training

In order to keep the HACCP System alive, it will be necessary for the company to promote HACCP on an ongoing basis. This can be done by building HACCP into the annual training programme, linking it with new training initiatives in hygiene or Statistical Process Control for example. Noticeboards and suggestion boxes can also be used to good effect, as can quizzes within company in-house newsletters. Another way of ensuring that the workforce remains aware of the system is to regularly report successes, failures, audit performance and changes down through the line management.

It is essential that the HACCP Team continually keep their awareness of new emerging hazards up to date. Again, this could be achieved by attending external seminars and reviewing literature. Membership of industrial food research associations and professional bodies can be particularly useful.

8.5.6 Design of new training material

Whether initially you did much of the HACCP training 'in-house' or not, you may wish to design internal training materials for future needs. This can be cost effective providing you have people who are suitable to act as trainers for the company. It is a complete waste of time and money to allow ineffective trainers to try to train people. A good investment in using your own training materials will be to train competent trainers. Don't make the mistake of using someone who happens to be available – consider the competencies needed for the role. These include being able to motivate others, communication and interpersonal skills, leadership skills, being able to manage diversity by recognizing and valuing differences in people, and finally having a sound, in-depth knowledge of the subject matter in which they are going to train others.

External 'Train the Trainer' courses are readily available and usually last a minimum of 2 days. It will be useful if the designated company trainer(s) have an input into the design of in-house materials. Use of an external consultant may also prove beneficial if resources are not sufficient within the company. Alternatively, you may be able to purchase off-the-shelf training packages which you can then adapt (Mortimore and Wallace, 1997). Computer-assisted learning is gaining in popularity and could also be considered.

Ongoing training activities should be seen as a way of continually raising company standards and as a way of ensuring that the HACCP System continues to grow.

8.6 Summary of maintenance requirements

You should now be in a position to pull together a summary of your HACCP Plan maintenance requirements. It may be helpful to formalize the proposed requirements at the HACCP Plan implementation stage and keep this document within the HACCP Plan. This is not an essential requirement of the HACCP Principle but it is of practical use for everyone involved to know the approach being taken. This, too, can then be audited for compliance.

The HACCP Team at the Iced Delights ice-cream factory came up with the example shown in Table 8.9.

Table 8.9 Ice-cream – HACCP System maintenance requirements

Maintenance requirements	Approved by: . . . . . A. . Jones. . . . . .
	HACCP Team Leader
	Date: 18.9.97

1. Quarterly audit of HACCP Plan
2. Annual plan revalidation
3. HACCP Plan to be revisited for all process ingredient changes
4. Quarterly CCP log sheet review for deviation trend analysis
5. Monthly review of customer complaint data for trends
6. 6-monthly simulation of trace/recall procedures
7. Ongoing technical information update through symposia and technical journals
8. Quarterly analysis of training needs and conduct refresher training of operators as required
9. Monthly HACCP Team meetings

9

Broader applications – linking HACCP with other Quality Management techniques

Manufacturers and caterers must consider many other requirements alongside HACCP in order to assure both the safety and quality of the food produced. These requirements make up the HACCP support network discussed in Chapter 3. Some of the prerequisite or support elements, such as Good Manufacturing Practices and Supplier Quality Assurance have been discussed in earlier chapters; here we will look primarily at the relationship between HACCP and other formal Quality Management Systems, such as ISO 9000 and Good Laboratory Practice (GLP) accreditation schemes.

Also within this chapter, we will consider the broader applications of the HACCP technique. Finally, we will take a look at how HACCP might be used as a development tool in a food business which does not have a basic HACCP support network in place.

9.1 HACCP and Quality Management Systems

First of all, what do we mean by a Quality Management System? Simply, all of the activities which go on in the company to ensure that it meets its quality objectives. In this respect HACCP itself can be considered as a Quality Management System (QMS) in that it is an activity which helps to ensure that the objective of manufacturing safe food is achieved.

Many companies base their QMS on the international standard series of BS (British Standard) EN (European Norm) ISO

(International Organization for Standardization) 9000 (ISO, 1994). The system can be formally accredited or the requirements used as a framework for an in-company system. The ISO 9000 standard can be, and is, used across a broad spectrum of activities in many organizations. Already, your company may be operating to the ISO 9000 standards, or alternatively you may view it as a goal to be reached within the next 1, 2 or even 5 years. Whatever the situation, you may have many questions regarding the relationship between HACCP and ISO 9000, and if you have neither, which should you progress towards first?

ISO 9000 is a Quality Management System aimed primarily at preventing and detecting any non-conformity during production and distribution of product to the customer, and by taking corrective and preventive action to ensure that the non-conformance does not occur again. ISO 9000 means that the product meets its specification 100% of the time. There is obviously a danger here in that if an unsafe product is specified, the Quality System will ensure that you make an unsafe product every time. This will also apply to the adherence to GMP standards – ISO 9000 will only serve to ensure that you operate to the GMP standards specified within your quality system. Clearly, this will include both legal and customer requirements, but, globally, there are still differences in legislative requirements and in what is interpreted as the minimum 'standard' where GMP is concerned.

How, then, can you ensure both that you specify a safe product and that you consistently meet the specification every time? The answer is to use HACCP, and to use the ISO 9000 approach to manage your HACCP System.

ISO 9000 and HACCP, concerned with Quality and Food Safety Management respectively, have much in common. Both systems require the involvement of all company employees, the approach taken is very structured and in both cases involves the determination and precise specification of key issues. Both systems are Quality Assurance Systems, designed to give maximum confidence that a specified acceptable level of quality/safety is being achieved at an economic cost. Quality Control techniques, i.e. statistically valid inspection and testing, are used as a vital part of the Quality Assurance System, to monitor that the control points – quality and safety – are being adhered to.

9.1.1 Using a Quality Management System to manage HACCP

In managing food safety the highest degree of confidence may be achieved by:

1. using a HACCP System which has been established by experts; and
2. ensuring that the HACCP System is maintained 100% of the time by using the ISO 9000 approach to meeting the specification (in HACCP terms, the Critical Control Points).

ISO 9000 is a series of standards which includes the requirement for 20 clauses to be implemented. The standard of ISO 9001 contains all 20, whilst ISO 9002 has 19 of the 20 and ISO 9003 only 16. Let's consider how each clause of the ISO 9001 standard can be applied to HACCP (Table 9.1).

Every one of the 20 clauses of ISO 9001 has relevance to HACCP and in many instances it is vital that HACCP is supported by such procedures. For example, HACCP can be a very effective way of managing food safety but only if:

- calibrated equipment is used;
- people are properly trained;
- documentation is controlled;
- non-conforming product is clearly identified and controlled;
- the system is regularly verified through internal audit; and so on.

A useful summary is shown in Figure 9.1.

In many cases, HACCP is not backed up by such disciplines and in these instances the company concerned may feel complacent in having a HACCP Plan, completely unaware that it may not be working effectively.

You certainly don't have to have a company quality system certified to ISO 9001 before you start HACCP, but you should be aware of the relationship between the two and how you can use the framework of ISO 9000 as a guideline for installing the procedures that will make your HACCP System secure. An additional bonus in taking this approach is that you will be well down the road towards ISO 9001 accreditation at a later date should you choose to do so. If you have neither HACCP nor a QMS, it is strongly recommended that, alongside the progression of HACCP, you use the QMS framework to support the HACCP implementation into the workplace, rather than do HACCP first followed by ISO 9000. The two are interlinked and have a real synergy. HACCP can also assist in focusing on the critical quality attributes (section 9.3.1), which can help to ensure a really well-targeted QMS.

Table 9.1 ISO 9001 – its application to HACCP

ISO 9001 clause	HACCP application
4.1 Management responsibility	– There should be a policy on HACCP Implementation which has been signed off at senior management level. A company which has a Quality Policy in place may include references to use of HACCP in managing food safety. – Responsibility and authority within the HACCP System must be clearly defined. A 'Management Representative' may be assigned to the HACCP System in a similar way to that of an ISO 9000 system. Frequently terms such as HACCP Co-ordinator are used for this role. – One of the requirements of a HACCP System is that its effectiveness should be verified. A regular review by management is recommended for HACCP and can be incorporated into Quality Review Meetings or done as a separate session; information such as complaints data, status of HACCP training, number of non-compliances at CCPs can be discussed and corrective action agreed.
4.2 Quality system	– HACCP itself is a Quality System but this clause specifically considers all activities which could impact on the 'Quality' of the product and ensures that they are consistently documented; a quality product is also a safe product. – Procedures for HACCP will be incorporated into the Quality System. Critical Control Point monitoring may be detailed within work instructions. – The HACCP approach can be extended to facilitate the identification of critical quality control points (QCPs) thus facilitating the development of a 'Quality Plan' which includes both Safety and Quality elements.

4.3 Contract review

Food safety is often assumed rather than referred to within a documented specification. The Contract Review clause of ISO 9000 considers the relationship between the customer and supplier. The customer is increasingly looking for assurance that food safety is being managed on a day-to-day basis. A combination of HACCP and ISO 9000 goes a long way towards providing this level of confidence.

It is useful to take the same approach within a Supplier Quality Assurance System where requirements should be clearly specified and raw materials which are Critical Control Points tightly controlled.

4.4 Design control

The HACCP process starts at the product concept stage when the inherent product safety should be considered. Designing food safety into a product is essential and food companies would be well advised to use ISO 9001 to ensure that 'critical' steps are controlled. Design Control will include a hazard analysis risk assessment in order to ensure that the product is designed safely. Design verification is performed to establish that the requirement for safety has been met.

4.5 Document and data control

– All HACCP documents need to be controlled by being reviewed, signed and dated by authorized personnel. This will prevent the use of out-of-date documents.

– Each HACCP document should ideally bear a unique reference number. This is useful for cross-referencing with CCP log sheets, for example.

– Food safety data should be maintained in an ordered manner to enable trends to be analysed and corrective and preventative action taken where necessary.

– When changes to the HACCP Plans are made, new issues must go to listed copy-holders and obsolete documents destroyed.

Table 9.1 *continued*

ISO 9001 clause	HACCP application
	– Ad hoc photocopying of 'uncontrolled' copies must be avoided and, where necessary, these copies should be clearly marked 'Uncontrolled'.
	– Control of artwork for packaging should be reviewed and signed off. This may be essential where 'Risk Communication' is the main method of control, e.g. with fish bones, fruit stones or allergens.
4.6 Purchasing	– Purchasing covers everything from raw materials to subcontractors. Written and agreed specifications should include food safety considerations.
	– Control of subcontractors should be managed through assessment and records retained. This will include calibration, process equipment servicing, hygiene, third-party laboratory services, laundry facilities and pest control.
	– Suppliers should be assessed on the basis of risk, to ensure their ability to deliver to the specified requirements. In the case of suppliers of 'high-risk' materials or services, this will include audit of the suppliers' premises.
4.7 Control of customer-supplied products	As this applies to product or ingredients supplied by the customer, it is important to ensure that these items are not left out of a HACCP Study. Specifications will be needed to ensure that the food safety information is available.
4.8 Product identification and traceability	– It is essential within the HACCP System to be able to trace batches of raw materials or products in the event of a failure at a CCP.
	– A written recall plan should be maintained in order to minimize the effect of any failure by being able to trace and withdraw any defective product.

4.9 Process control

HACCP is underpinned by working to Good Manufacturing Practices as a daily routine. Compliance with GMP Codes of Practice and reference standards falls under this clause. The following key areas should be included:

(i) Buildings – all facilities from raw-material storage areas, through process and despatch. Including employee amenities.

(ii) Plant and equipment – process capability, preventative maintenance, hygienic design and cleaning, routine process parameters.

(iii) Personnel – training, health screening.

(iv) Cross-contamination – at all stages where this could be a risk.

(v) Waste materials – clearly identified, segregated and disposed of in a safe and hygienic manner.

(vi) Environmental control – atmosphere, ground water.

4.10 Inspection and testing

– Where raw materials are CCPs, these ideally should not be used until confirmation of conformance to agreed specification has been received.

– Materials used before being certified as meeting specifications should be traceable in order to allow a recall at a later stage if necessary.

– Where possible finished products should be held until confirmation of all CCPs having been met in full has been received.

– Records of all inspection and test results should be maintained in a manner that will allow trend analysis.

– Personnel carrying out testing should be trained and qualified appropriately. They should also be assessed as being capable for the job.

Table 9.1 *continued*

ISO 9001 clause	HACCP application
4.11 Control of inspection, measuring and test equipment	– Effective control of CCPs relies on accurate measuring methods or equipment. All equipment needed to monitor a CCP should be of known accuracy and precision and calibrated on a regular basis.
	– Equipment should be capable of accurately measuring the inspection and test criteria as laid down in the HACCP Plan.
	– All measuring equipment should be status marked so as to make it clear to all personnel what is calibrated and what is to be used for general guidance only.
	– Equipment must be maintained and stored correctly between calibrations in order to avoid damage.
	– Records of calibrations should be kept.
4.12 Inspection and test status	– There should be a clearly defined method for identification of the inspection and test status of any raw material, product or equipment to prevent its being used inadvertently.
	– Any product which does not meet specification should be released only through authorized concession and discussion with the customer. Personnel involved in making these decisions should be qualified to do so.
4.13 Control of non-confirming product	– The HACCP Control Chart will define who is responsible for taking corrective action in the event of a deviation. This must include what to do with product made while the system was out of control, i.e. to rework, recycle or reject.
	– Procedures must be developed to ensure that all non-conformities at a CCP are recorded. This will enable trends to be analysed.

ISO 9001 clause	HACCP application
	– Once any non-conforming product has been brought back into specification it must be tested again to confirm conformance.
4.14 Corrective and preventative action	– HACCP is based on identifying where preventative measures are critical within the process in order to maintain control.
	– If a CCP is out of control, the underlying cause must be identified in order for the problem to be resolved permanently and not repeated. Actions should be such that a recurrence is prevented.
	– Corrective actions also include pest control audits and hygiene audits.
	– The corrective action taken in the event of any problem arising must be the right corrective action – new hazards could arise if the wrong corrective action is taken. An example of this is to shroud a piece of equipment in polythene because the roof above is leaking.
	– The effectiveness of any corrective action must be verified for confirmation.
	– All corrective actions must be recorded.
4.15 Handling, storage, packaging, preservation and delivery	– Hazards may arise from improper handling and storage of a product or raw material; also, if the packaging is unsuitable – unable to withstand distribution or handling by the customer. Areas for consideration include: • food contact packaging • control of artwork (usage instructions, ingredients and nutritional data) • storage and distribution temperatures • shelf-life

Table 9.1 *continued*

ISO 9001 clause	HACCP application
	• contamination risks • environment and building fabric • hygiene and pest control.
4.16 Control of Quality Records	– HACCP records will need to be retained in a controlled manner. They may form part of a due diligence defence or be needed during a regulatory inspection as demonstration of the effective management of food safety.
	– Quality-related records include product and raw material specifications, the HACCP Plans, Process Control records including CCP log sheets, calibration records, minutes of HACCP/food safety meetings, audits, training records.
	– Retention time must reflect both statutory regulations and product shelf-life: 3 years as a minimum for those records that demonstrate system management.
4.17 Internal Quality Audits	– Process Flow Diagrams must be audited as part of the verification. It is essential that these are complete and an accurate representation of the production process.
	– The HACCP System itself should be regularly audited by members of the HACCP Team in order to assess whether it is working, correct and applicable. Internal audits can act as a health check. Documented non-compliances raised must be corrected, allowing continued improvement of the HACCP System. The audit will be what makes the HACCP System live within the company rather than being a set of documentation on the shelf of the QA Manager's office.
	– Auditors must be trained and independent of the department being audited, and records of audits must be kept.

4.18 Training

– Effective HACCP relies on participation of knowledgeable and trained people across a wide range of disciplines. Skills in areas such as Hazard Analysis and Risk Assessment will be needed.

– HACCP and Food Safety Management will continue to develop. It is important therefore that future training needs are considered on an ongoing basis. Areas for inclusion may reflect system changes, emerging food safety issues, new processes and corrective action skills.

– Training records should be available for all members of staff and include all types of training, whether in-house, external, skills or awareness activities.

4.19 Servicing

– Often not considered relevant in manufacturing, but suppliers of vending equipment will be particularly aware of this clause as it requires the servicing of such equipment to be documented and controlled. Such equipment can be a source of hazards if not cleaned and maintained correctly. This needs to be considered in line with the servicing frequency.

4.20 Statistical techniques

– The construction of the HACCP Control Chart requires sampling regimes at each CCP to be documented.

– If decisions of conformance are being based upon results of sampling and testing, then it must be ensured that schemes are mathematically sound, i.e. the use of statistical sampling plans.

– Other relevant statistical techniques include process capability assessment and Statistical Process Control during CCP monitoring.

9.2 Laboratory accreditation

9.2.1 What is laboratory accreditation?

Laboratory accreditation is the systematic assessment and validation of a laboratory operation against a specific laboratory quality standard. It is normally carried out by an independent accreditation body and assesses the laboratory operation against a number of key elements, as defined in a laboratory quality standard. The quality standard may itself be based on the international standard 'General Requirements for the Technical Competence of Testing Laboratories' (ISO Guide 25) or the European standard 'General Criteria for the Operation of Testing Laboratories' (EN45001; EURACHEM/WELAC, 1993). The exact wording of laboratory quality standards will differ in different countries and in different schemes, but the key elements normally cover the following areas:

- organization and management of quality systems;
- audit and review;
- laboratory design and hygiene;
- sample handling;
- equipment;
- calibration;
- methods of analysis;
- quality control;
- records and reports.

9.2.2 Why is laboratory accreditation important to HACCP?

The laboratory operation is a critical part of any quality system supplying information to verify that products are within specification. The results of laboratory analysis are particularly important when they are being used to monitor or verify the operation of CCPs, and thus product safety. Here, it is on the basis of results that decisions are made and actions are taken, so it is vital that they cannot be disputed.

In the case of CCP management it is essential that not only are the laboratory-based tests accredited but also any analytical testing which is carried out on the production line or in the production areas. These should be included in the scope of the accreditation.

Laboratory accreditation is also important where the company plans to use its HACCP System as part of a defence in any litigation case. In this case, the company would need to provide evidence that its HACCP System was operating under control, and that monitoring and verification was being carried out using assured methods. Independent laboratory accreditation gives confidence in the laboratory operation and helps to support the HACCP System and the litigation defence.

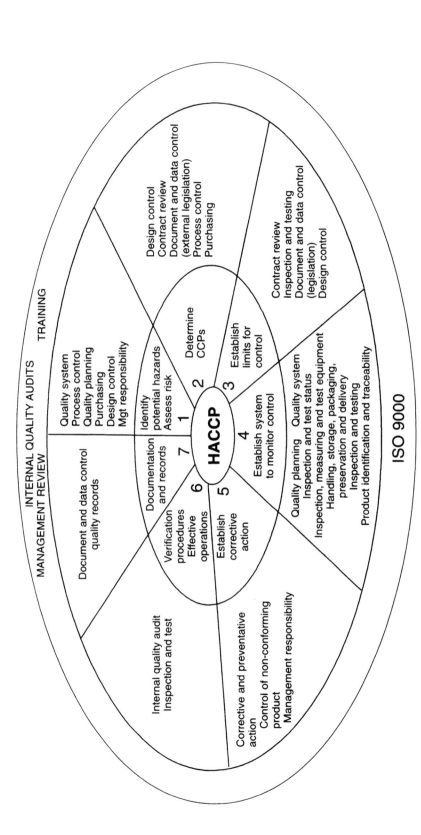

Figure 9.1 HACCP and ISO 9000 relationship diagram (reproduced with kind permission from *Guidelines for the Use of ISO 9001: 1994 in the Design and Manufacture of Food and Drink*, Lloyd's Register Quality Assurance, 1995).

There are many general benefits of laboratory accreditation. Specific benefits to HACCP and product safety are:

- That decisions and action are based on valid results.
- There is confidence that product safety specifications are being met.
- There is assurance that results are accurate and reliable.
- Accreditation will support a defence under litigation procedures.

9.2.3 Is accreditation necessary?

There is a strong case for laboratory accreditation in any organization but particularly in one operating a HACCP System for product safety. It is absolutely crucial that the results of all monitoring procedures at CCPs are irrefutable and can be trusted to demonstrate that the system is under control, or can be used as the basis for corrective action decisions. Laboratory accreditation gives confidence in the accuracy of results and an independent system lends support to any necessary defence under litigation procedures.

9.2.4 Benefits of the combined system – the integrated approach

The main benefit of using the combined system of HACCP, ISO 9000 and an accredited laboratory is confidence; confidence in ensuring that the CCPs are correctly identified and maintained, that the documentation is controlled and, in using both HACCP and ISO 9000, that 100% of the time, your product is safe. The laboratory accreditation ensures that test results are accurate and can be relied upon. Statistically based sampling schedules must be valid in order to meet the requirements of ISO 9000.

Having external independent assessment of the Quality Management and laboratory systems will ensure even greater confidence. To make the system really secure, an independently certified HACCP System would be the final element. This can be done to some extent by including HACCP in the scope of your ISO 9000 system and ensuring that the auditors are competent in both HACCP and audit skills. Third-party HACCP certification audits are becoming increasingly available and are usually carried out by research associations and inspection bodies. In some cases, specialist schemes incorporate both the elements of ISO 9000 and HACCP together with hygiene, e.g. HACCP 9000 (NSF, 1996).

9.3 Broader applications of HACCP

HACCP was originally developed as a technique for the identification and control of food safety hazards, and most companies will still want to use it to focus primarily on safety. However, the HACCP technique that you have learned in this book can also be applied to other areas beyond food safety, but a few words of warning . . .

Safety should always be the primary objective, and using HACCP for other issues should not be allowed to complicate matters and cause confusion such that product safety management becomes muddled and ineffective. In order to achieve this, it is best to use the HACCP approach separately in each of those areas and then manage the resulting systems through an integrated Quality Management System.

9.3.1 Identification of Process Control Points

The hazard analysis approach can highlight 'hazards' at many stages of the process which may not result in direct product safety issues if not controlled. There may still be safety hazards arising at a step which has no CCP because of a CCP later in the process, for example the risk of metal in several incoming ingredients where there is a metal detector at the intake stage and which will also be controlled by the end product's metal detection. As discussed in Chapter 4, a control point such as this can be managed as a 'Process' or 'Preventative' Control Point, in that early identification of the presence of the hazard will be more economical before the contaminated ingredients have been converted into expensive finished products. However, the final control point, in this example at the final metal detector step, is the critical one.

Other process control points will include quality issues (e.g. viscosity, label orientation, or the product's sensory attributes), which need to be controlled for a high-quality product. These so-called non-food safety 'hazards' may either be identified during brainstorming in a safety HACCP, where they would not be directly associated with CCPs, or they could be identified as part of a completely separate exercise focusing on quality. While these points may be critical to product quality, it is extremely unwise to call them CCPs as this causes direct confusion with the safety management system. Instead, these points are often referred to as 'Process', 'Manufacturing' or 'Quality' Control Points (QCPs).

Using the HACCP techniques independently to capture other Control Points ensures that they are secured in a systematic manner. As for safety HACCP, the terms of reference for each study need to be clearly defined, and then it is simply a question of following through the hazard analysis and deciding where the elements of control should be situated. In this case, use of a 'CCP' decision tree is unlikely to be appropriate as you may want to have process control points in a number of areas, not just at the final critical point, but the structured process of Hazard Analysis can be helpful, for example, if 'hazard' is redefined as a quality defect. In the same way as safety HACCP, the monitoring and control parameters should be established and these could be encompassed in the product specification or in a separate Process Control Plan.

If the HACCP Team identifies quality issues during a safety HACCP, it is important to the company's business that these are not lost. A useful and practical way of documenting these points is to use a HACCP and Quality Control Chart. This enables the differences between the food safety control points (the CCPs) to be very clearly and separately identified. An example is given in Table 9.2.

9.3.2 Finding legal control points

Here you can use identical procedures to those that we have described for safety HACCP but will need to redefine 'hazard' as a non-compliance with legal requirements.

This approach can be particularly helpful when wanting to distinguish between the true food safety controls and those which are a legal requirement. Sometimes a CCP will also be a Legal Control Point (LCP), for example in the requirement for pasteurized milk to achieve a defined time and temperature during the pasteurization process. However, some food safety legislation goes beyond the control of hazards, for example the Food Safety Act in the UK (1990) where it is an offence for the foodstuffs to contain non-hazardous foreign material, being 'not of the nature, substance or quality demanded' by the consumer. Some companies choose to manage all foreign material contamination as a CCP, which can be confusing for operatives trying to understand the difference between a contaminated food that is safe to eat and one that is not. Managing these types of issues as LCPs can give better clarity to the overall system and aid communication to the workforce. LCPs can be documented using the HACCP and Quality Control Chart approach (Table 9.2).

For the more computer literate, the chart could be drawn up as a spreadsheet with separate data fields for the CCPs, QCPs and LCPs. You could then choose to view (or print off) separate charts, depending on the purpose, i.e. the HACCP control chart might show only the CCPs, a Legal Control Chart the LCPs, and a Quality Control Chart within the ISO 9000 or QMS system might show all three elements.

9.3.3 Prediction of product spoilage

The hazard analysis techniques can also be used to predict product spoilage. By establishing appropriate terms of reference this can be used to set an achievable shelf-life.

For microbiological spoilage, in the same way as for safety, you

Table 9.2 HACCP and Quality Control Chart

HACCP & Quality Control Chart								Date:	Supersedes:
								Approved by:	Ref No.:
Process Step	Control Point			Control Measure	Critical Limits	Monitoring		Corrective Action	Responsibility
	CCP	QCP	LCP			Procedure	Frequency		

CCP = Critical Control Point = food safety;
PCP = Process Control Point = quality other than food safety; also preventative controls.
LCP = Legal Control Point = legal compliance.

can use hazard analysis to establish which spoilage organisms are likely to be present in the raw materials, and which might survive the process, or which might cross-contaminate from the process environment. You then need to use knowledge of growth rates of these organisms in similar situations (e.g. products with similar intrinsic factors and the product storage temperature) to predict the likely achievable life. This will normally need to be backed up with shelf-life experiments, where the product is examined for the micro-organisms of concern, or by challenge studies, where organisms are inoculated directly into the product and their growth/survival potential is evaluated.

The hazard analysis technique along with knowledge of chemical and physical reactions in food can be used to give predictions of changes, such as in flavour, physical appearance (e.g. colour, texture) and in product taste. The potential for leaching of taints from other product components or from packaging during the shelf-life can also be evaluated, but of course if this is a safety issue it should be picked up in the main safety HACCP Study. Again, this will normally need to be backed up with shelf-life experiments, this time of an organoleptic or analytical nature, but an advantage of using the hazard analysis technique is that confirmatory experiments can be targeted accurately at essential information, thus saving on resources required for a full experimental trial. Product spoilage controls can be managed as PCPs.

9.3.4 Other applications

(a) Evaluation of cleaning procedures

An effective cleaning programme as part of GMP is an essential prerequisite within any food processing environment. However, cleaning procedures may also be considered as part of the main HACCP Study, where they will often be critical in preventing serious microbiological contamination. Additionally, cleaning itself may cause a chemical or physical hazard if not properly controlled.

It is important to identify where cross-contamination through poor cleaning can occur. Because of this, it is often helpful to conduct a separate HACCP Study of the cleaning procedures which may result in specific CCPs being established and added to the HACCP Plan. Prevention of allergen cross-contamination, for example, may require that critical controls are identified within the cleaning schedule.

(b) Designing preventative maintenance schedules

In the HACCP Study we determine where the equipment and environment are critical to product safety. Carrying out an additional hazard analysis on the equipment operation itself may highlight where it is likely to fail. These two activities will help to establish an effective preventative maintenance programme.

(c) Application to plant design

Hazard analysis techniques can be used at the concept stage of equipment and factory design to ensure that plant, equipment and factory do not cause food safety hazards through inability to clean, or through the risk of physical contamination.

(d) Prediction of malicious tampering opportunities

Use the Process Flow Diagram and hazard analysis technique to determine where tampering opportunities exist. Consider the following points:

1. Where might the product be accessed easily?
2. Where in production is it open to the environment?
3. Be aware of situations where operators may feel begrudged.
4. Is the packaging resistant to access and will it indicate if tampering has occurred?

(e) Assessments for health and safety

Both HACCP and Health and Safety systems use a risk assessment approach. The 'hazard' is defined, this time as a health and safety issue such as repetitive strain injury, fire, explosion, or forklift truck accidents. Both the process itself and the control of substances harmful to health will need to be considered in order to assure operator safety and well being. Hazardous chemical control also plays a role in HACCP, mainly through the clear identification of chemicals, prevention of cross-contamination and awareness of the types of chemicals present on site. Occupational health procedures such as pre-employment medical screening and return to work procedures may be an essential prerequisite in a high-risk operation. HACCP and Health and Safety can be linked together through the sharing of work instructions at operator level.

(f) Environmental management

Structured environmental management systems are now available, (BS7750 is the British Standard and ISO 14000 the International Standard) and increasingly this area is being regulated on an international basis. One of the most obvious links into HACCP is the emphasis on recycling, which can create additional hazards. With

Figure 9.2 'Prediction of malicious tampering opportunities'.

recycled paper and card, for example, it is important to consider the increased likelihood of metal staples within the material itself; also the full chemical breakdown of food-contact packaging may not be known if traceability is poor.

(g) Incident management and recall procedures

If the HACCP System fails or another unforeseen crisis occurs, e.g. to the factory building through fire or explosion, it will need to be managed in a controlled, systematic way, in order to minimize the damage to consumers and the business itself. Personnel who are trained in HACCP will be familiar with the identification and analysis of food safety hazards. This will be helpful during the management of an incident and also before this when putting an incident management system in place, in readiness for such an occurrence. One of the initial phases involves undertaking a business risk assessment. This normally involves a series of questions:

1. What are the potential risks to our business?
2. What would be the outcome?, and
3. What is the likelihood of the risk being realized?

This process of business risk assessment tends to be carried out systematically and some companies choose to use a scoring method for recording their output.

It is vitally important to be able to trace effectively any potentially unsafe product, as with product-related incidents there is a strong chance that the product will need to be recalled from the market place. It is difficult, if not impossible, to ensure a 100% return of product, and the best option may be to use the media in order to inform the public. Trained personnel, with media-handling skills, will be a real advantage in reassuring your consumers that the situation is under control.

9.3.5 Application of HACCP to non-food products

As previously stated, HACCP was originally designed for use in food safety and has traditionally been used in that area. However, the techniques are equally applicable to non-food products, as might be expected since HACCP had its roots in the engineering system, FMEA.

Some of the products in the non-foods area are similar to foods in that their safety is directly related to application to and by the consumer, e.g. shampoo or soap which must be safe for direct skin contact, or washing powders which must not leave any residues on fabric which would harm the skin. Other products, such as electrical goods, have a much longer usage life and it is important to ensure that they are not only safe for use initially, but also that they cannot develop any faults which will make them potentially unsafe later in the product life.

Other products may have a potential effect on food, e.g. clingfilm, foils, cooking utensils or food-storage containers. These must not pass any harmful substances into the food products they contact. Many countries have legislation covering specific substances and migration limits, and the hazard analysis techniques can be used to highlight those substances that are likely to be present, along with how they are likely to react under the expected use conditions. For example, if heat is applied to a plastic microwave cooking utensil, will any substances pass into the food which is being cooked, and if so will this be harmful to the consumer?

Safety in the non-foods area is further complicated by the fact that some products are unsafe by their very nature, e.g. household bleach. With products such as this the emphasis is placed on instructions for safe use and safe packaging which will neither burst allowing product to escape, nor be accessible to children or individuals who may not understand the safety instructions. Products such as aerosols may be explosive or inflammable, in which case safety instructions are again critical.

The HACCP techniques can be applied to these product areas in exactly the same way as to foodstuffs. The members of the HACCP Team will of course be expert in the particular product area under study, and they will need to consider hazards associated directly with the use of the product, along with indirect or delayed safety hazards. When the hazards have been identified and the risks analysed, CCPs can be established as for food products, and these will require specific control criteria and monitoring systems to be implemented. The actual HACCP Plan should be managed in exactly the same way as we have seen in earlier chapters of this book.

It is also possible to use HACCP techniques to identify process control issues for non-food products in the same way as we saw in section 9.3.1 for foods. Here again it is important not to confuse safety with process control.

9.4 Using HACCP in targeting resource – a strategy for continuous improvement

If you have read Chapters 1 to 8 before beginning Chapter 9, you will by now have a good understanding of what you need to do to implement and maintain a HACCP System. If you are starting with nothing in place, you may be feeling overwhelmed by what you have to do and be somewhat despondent (Figure 9.3). Some companies, particularly smaller ones, often think that use of systems such as HACCP are for the large, sophisticated manufacturing companies with plentiful resources. Other companies, on considering HACCP and being made more aware of hazards and risks, realize that an effective HACCP System relies on firm foundations and prerequisite programmes which they perhaps do not have. As we saw in Chapter 3 the foundations may include:

- Working to Good Manufacturing Practice (GMP), including documented and verified cleaning programmes.
- Having an established Supplier Quality Assurance System (SQA).
- Using Statistical Process Control techniques (SPC) to assure process capability and process control.
- Food safety and quality-related training programmes which are linked into business strategy – both skills and awareness training.
- Having a fully developed Product Recall and Incident or Crisis

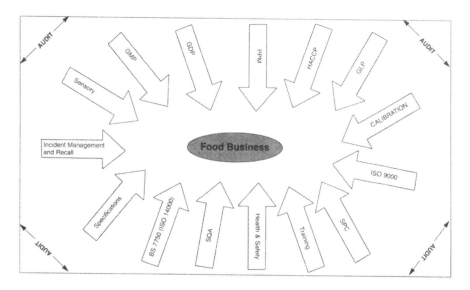

Figure 9.3 The broader perspective – where to begin?

Management Plan in place which is regularly challenged, supported by accurate and complete traceability of raw materials through to distribution to the customer.
- Regular use of problem-solving techniques such as brainstorming, which we have covered previously; also other techniques such as Cause and Effect to tackle recurring problems in a way that will ensure that they are permanently resolved, i.e. treating the cause and not the symptoms through daily fire-fighting.
- Having all quality- and safety-related procedures documented.
- Use of a Quality Management System (ISO 9000 approach) to pull all of the above activities together.
- Calibration of key equipment.

In addition to this, the management team may be faced with trying to implement systems in other areas such as Health and Safety, and Engineering. This may feel like an impossible task and hard enough to tackle even with full commitment from the team.

However, perhaps the culture of your company is not quite as it should be – some managers and staff may still be content with the way things have been in the past, and there may be no perception of the need to change to a new way of working.

If you don't have any of the above list of activities in place and want to use HACCP – don't worry, HACCP is the best place to begin. HACCP can be used by everyone and there is no right or wrong time to start using it. The normal use of HACCP is when hygiene is under control and it is fair to say that this view is widely accepted. A common misconception is that if you don't have any written specifications and procedures, and have very poor GMP and hygiene, then you are in no position to use HACCP. To the authors, it seems only common sense to say – use HACCP now to help you decide where to begin, i.e. in prioritizing against food safety risk and targeting resource.

Hazard analysis of your processing operation will enable you to focus on high-priority areas for improvement and draw up a realistic action plan. To do this you will need to draw an accurate Process Flow Diagram and, as we saw in Chapter 3, a benchmark audit at this stage is also invaluable in helping to establish where you are in relation to the end goal. Calibration of your key process equipment, such as that which controls and monitors temperature process steps, will also be done at an early stage, but this will be identified during the baseline audit. These activities can be done fairly easily in any size of business, whatever the level of maturity. Other than that, you need to do very little before starting the HACCP study. Once you have the Process Flow Diagram you can carry out a hazard analysis exercise in the normal way. If you have no raw material specifications, use known data from reference books, a hazard database, or bring in a HACCP expert to help you. Once you have identified the hazard (remember to be as specific as possible) you can determine what the control measure **should** be, rather than simply recording what your process operation currently uses. Add an additional column then to the Hazard Analysis Chart to indicate whether the control measure is currently present (Table 9.3) and another three columns to record your action list, who will be responsible for making it happen and the time scale.

The appearance of a 'No' response in the first column will indicate that corrective action is needed and that the control measure must be put in place. How should you prioritize and set the time scales – after all, there may be a long list of actions required? Use the CCP Decision Trees to help you in targeting critical areas requiring control. This may perhaps seem an obvious answer but is particularly helpful where capital expenditure is involved.

The raw material CCP Decision Trees will be helpful in providing focus to SQA programmes. This can be taking place at the same time as a production team is drawing up the Process Flow Diagrams and should involve those personnel responsible for purchasing. Many companies purchase hundreds of raw materials (ingredients and packaging). If you have no specifications, a blank pro forma can be sent to all suppliers for completion but using the raw material decision tree will help you to see where Certificates of Analysis are

Table 9.3 Are all the required control measures in place?

Process step	Hazard	Control measure?	Currently in place? Yes/No	Action required	Responsibility for action points	Time scale
Incoming materials chocolate chips	*Salmonella*	– Agreed specification	No	Obtain specification from supplier. Agree content and sign off	QA Manager	1 month
		– Certificate of analysis from supplier	Yes	Check GLP status of supplier's laboratory	QA Manager	1 month
		– SQA visit to supplier	No	Conduct an audit of the supplier	QA Manager	6 months
Weighing: 1 Bagged sugar 2 Chocolate chips	– Ingress of paper from sack into sugar or other foreign bodies	Sieving through 8 mm mesh sieve	No	Purchase a sieve for raw material, weigh-up room	Engineering Manager	1 month
	– Pathogen cross-contamination from operative	Hygiene training	No	Identify the priority staff for training. Call in an external trainer to deliver a training programme	Human Resources Manager	12 months

needed and which suppliers should be audited as a priority in order to assess their level of competence. It is important that this activity happens as quickly as possible because you might want to request that your suppliers of high-risk raw materials also use HACCP. They can be working on their HACCP Systems while you are working on yours.

In using hazard analysis and the CCP Decision Trees for both raw materials and process steps, you will also be able to identify training needs. Training can be costly, particularly if done externally, and priorities will need to be assessed. You will have identified where basic training in food hygiene is important and, if this involves a large number of people, it may be cost effective for you to conduct this in-house. HACCP, SQA and auditor training may also be needed. By identifying how many suppliers need auditing, you will know what level of resource is needed – whether it can be done by external consultants, whether several in-house QA auditors will be needed, or whether the QA Manager will be able to do it.

Use of an established technique such as HACCP early in the development of a company can do a lot in helping to gain commitment for change from all staff. HACCP is common sense; if food safety hazards are identified then they must be managed. HACCP can be used to develop a strategy for improvement which is based on a real need to manage food safety.

10

Epilogue

10.1 Introduction

At the end of a specialist book such as this it is sometimes possible to lose sight of everything that was going on around you before you picked it up and started reading. Well, back to the real world now and, if you have not yet begun to plan how to apply and implement the HACCP Principles, then this is where to start. Before we leave you to your task, we wanted to share some of our thoughts on food safety management in the future. You will want to put in a system that will have a 'long shelf-life', one that is capable of adapting to encompass new issues as they occur. And, as the industry changes over time, hopefully your food business and its food safety control will move with it.

10.1.1 Drivers of change

First, let us briefly consider what the main drivers of change in the food industry have been over the past 50 years or so, and what may develop in the future. By doing so we can explore what we may need to put in place to ensure a safe food supply for the future.

(a) Legislative

In general terms, regulatory requirements in many countries have contributed to the change in food industry practices over the years. It could be argued, for example, that the UK Food Safety Act (1990) and its obligation for food business proprietors to take all reasonable precautions in the manufacture and supply of safe food has largely been responsible for the surge in supplier auditing. In the

USA the new requirement for regulated HACCP in the meat and seafood industries is, no doubt, acting as a catalyst for review and update of existing HACCP Plans in those industries. The challenge for governments worldwide is to impose reasonable frameworks within which the issues of food safety can be managed. These frameworks must encompass education as well as manufacture, and encompass systems as well as standards.

(b) Environmental change

This encompasses not just the physical world but also the microbiological world which allows mutagenesis of new strains of organisms. Emergence of new pathogens in recent years has led to the review of existing control measures. For example, the methods for limitation of cross-contamination with *E. coli* O157, or the prevention of the likely BSE causative agents from entering the human food chain, are areas still under research. The mutagenicity of micro-organisms will become of increasing concern with the appearance of more chemical- and antibiotic-resistant strains of increasing virulence. Ultimately the approach to food safety will necessarily be based on prevention of contamination rather than preservation or cure. If we have robust HACCP Systems in place, then this will surely contribute to prevention of, as yet unknown, hazards. However, we will need to review the continued effectiveness of existing control measures constantly. Maintenance of the HACCP System is, clearly, vital if we are to rely on HACCP for food safety.

(c) Sociological change

Internationally, consumers have now had the opportunity, through travel and media, to experience a wider culinary culture. In addition, people have increasingly migrated from country to country around the world, taking their food cultures with them. Food influences from around the world are now common in all domestic households and television cookery programmes are constantly urging people to expand their horizons even more. In today's mobile society, eating out is now very much more common and there is a decline in the family meal occasion where the family cook prepares a meal for the whole family to eat together. Individuals in the household are now far more likely to prepare a meal just for themselves. This means that food-handling skills and the preparation instructions on packaging are needed by a much wider set of people. The provision of food out of home has also moved away from traditional restauranting, where cooks would learn their skills under a hierarchy from the head chef.

Much of the food eaten outside of the home is through fast-food and on-the-move catering outlets. These require a larger number of people to be involved in the preparation of food and in flatter, less knowledge-based structures. Turnover of staff can be high in these businesses, so the use of HACCP to identify Critical Control Points will continue to offer a focused and practical way forward.

(d) Food distribution

The way that food reaches consumers through the supply chain, the routes to market, has altered considerably and this has brought new hazards with it. The shape of food retailing has changed dramatically; in some countries, notably the UK, but also western Europe, North America and Australia, the growth of major grocery retailers has been phenomenal. In the UK 80% of the value of all grocery purchases is vested in only five retailers, each of which has several hundred stores across the country. Similar scenarios are developing elsewhere. This poses a number of issues. The food supply chain has become so pre-packaged and sanitized that very few people in these stores actually handle food, they are essentially involved in merchandising. The number of people in retailing at store level who are skilled at food handling and preparation, with the exception of a few people on service counters and in-store processing, has decreased. The only food safety issues at large here are generally ones of temperature control and shelf-life.

The ability of these stores to source globally and to offer a wide choice of ethnic foods has succeeded in making consumers' tastes more eclectic. For the stores themselves, their contact with a global supply base gives them an unique opportunity to further the use of HACCP in a consistent way. Perhaps they no longer have the resources to visit such a wide base of suppliers as frequently as they would if they were in the same geographical region. Therefore, they must rely on systems such as HACCP and supporting prerequisites being understood and managed in the same way by suppliers based in Manchester (UK) or Madagascar.

(e) Food processing and technology

Producers and processors have not been insulated from change during this time. Primary producers have seen the advent of intensive agriculture and factory farming and have not escaped unscathed from the mass use of chemicals and intensive agricultural practices. The structure of manufacturing organizations has been affected by merger and rationalization. Large companies now source their ingredients on a global basis for reasons of economy and supply through logistically controlled distribution systems to fewer customers. The control of food safety in these companies is now vested in the hands of a few experts. The challenge is to create the right amount of awareness and skills at the critical food contact points through education and training, and then to maintain this at adequate levels. Throughout all this change the food industry has introduced new controls and, perhaps in some instances, these were perceived by consumers as coming too late – BSE is a good example here.

What next, then, for the future? How might our HACCP and food safety management systems be improved to ensure that new developments, whether technological, sociological or whatever, are managed properly such that food safety incidents are prevented? Two main themes occurred to us as we started to write this epilogue. First, a more closely integrated food safety system across the supply chain, and secondly, more effective training and education.

We will look at each of these areas in turn and consider how HACCP may be used to make a positive contribution

10.2 Integrated food safety management system

What we see for the future is a fully operational matrix of activity across the supply chain. Shared goals through a mutual desire to improve food safety management should be a common theme – in many cases the drivers for a primary producer will be the same as for a processor, caterer or retailer. In some instances one may be a driver of the other, for example retailers driving improvements through their suppliers in the processing sector and processors demanding improvements of their primary producers. SQA systems clearly affect the entire chain, but what about the opportunity for shared hazard analysis, problem solving, quality systems linkages and audit activities? Figure 10.1 suggests how this might work.

A commonality of approach for the implementation and support of HACCP can serve as a vehicle for the integration of the associated prerequisite and other Quality Management Systems. This will

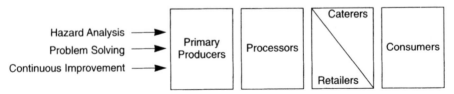

Figure 10.1 Joint food safety programmes: crossing supply chain barriers.

provide the foundation for knowledge transfer and will allow, for example, hazards to be jointly identified by different parts of the supply chain. Where they arise in one part of the supply chain they may actually be controlled in another part. Such integration of systems must occur in order to enable effective control to be imposed.

In the future we could envisage a knowledge transfer system, perhaps electronic, which spans the supply chain. Use of a product safety assessment, as outlined in Chapter 5, could facilitate the handing on of the essential hazard analysis information from one area of the supply chain to the next – a sort of 'passport' system.

With developing and more readily accessible technology, use of electronic data transfer will increase; perhaps, also to the consumer who will be able to use a scanner in the retail outlet not only for identity and price information but also to read product safety data. Some superstores are already beginning to offer this scanning facility in-store so that consumers can access a database of nutritional information and details of potential allergens. Eventually, perhaps, the use of scanners could extend to the home environment with the product information encoded in the label.

10.3 Education and training

In addition to the integrated food safety management approach, the increased regulation of HACCP, global sourcing and world trade agreements will have a major impact on the training and education needs across the food supply chain.

Already in the past few years there has been a large increase in the demand for HACCP training. Unfortunately, to date, although some HACCP training has been delivered by experienced HACCP trainers, there is also a considerable variability in the standards of training available, given the lack of regulation or standardization in this area. In many instances HACCP training is being offered by consultant training providers as a range extension to their hygiene training courses, yet many fail to make the conceptual leap from hygiene management to hazard analysis, risk assessment and control (Mortimore and Smith, 1998). This is a problem largely facing the industry at present: many HACCP experts are skilled presenters yet few have a true education in learning theory and training skills; on the other hand, training experts are rarely HACCP practitioners. Training in HACCP, then, can often end up leaving the trainees confused and with a superficial theoretical knowledge. In the UK, the RIPHH facilitated the development of a HACCP training standard (1995b), and almost more important than this went on to certify HACCP training courses against that standard, offer examinations as verification of learning, and to register training centres and trainers. Trainers have to demonstrate both HACCP experience and training skills. This is a good model and has helped to focus both the trainer and trainee towards their joint objective of successful knowledge transfer during the training intervention.

Training in HACCP, currently, tends to be just that – fairly narrow in scope – when you consider all the other aspects that are part of food safety management. While the demand for HACCP training is likely to increase, hopefully it will also broaden out to include the other elements of food safety management such as SQA, GMP and Quality Management Systems knowledge. This will then look more like a total food safety management programme with HACCP as the core element.

Earlier, we touched on the need for increased consumer learning and, in part, this can be provided through the introduction of food safety knowledge into the national curriculum of schools worldwide. This alone will not be sufficient – what is needed is a coherent and co-ordinated approach to the education of consumers. In the UK the AIDS awareness campaign was extremely successful in targeting leaflets to individual homes, hospitals and doctors' surgeries, schools' programmes and TV advertising campaigns. If the same approach were taken for food safety management promoting a HACCP approach for producers, caterers, retailers *and* consumers alike, we would all be able to work together to reduce the level of foodborne illness.

Many of these issues may affect your business, if not directly as described, then in some variation on the theme, and there may be other issues which we have not thought of. Whatever arises, it is important that your, and every other, business is prepared so that we are able to reduce the level of foodborne illness and control any new hazards that may occur in the future. Proper use of HACCP will go a long way towards preparing for these eventualities.

10.4 Closing thoughts from the authors

Finally, a few thoughts to bear in mind if you are embarking on your initial HACCP endeavours:

- Keep it simple and focused.
- Be clear on your objectives.
- Choose the right people for the job and train them properly.
- Ensure that your HACCP Team members know what is expected of them.
- Prepare thoroughly – to fail to prepare is to prepare to fail!
- Do not make assumptions.

- Work in an organized manner.
- Make sure that all details are recorded, from who was on the HACCP Team to notes on the thought process during hazard analysis and CCP identification.
- Challenge existing beliefs – make sure that you have evidence of what is actually happening.
- Challenge current practices – are they really acceptable?
- Resist the temptation to make the HACCP findings fit the existing control and monitoring schedules.

We would like to wish you luck in applying HACCP to your operation. We have enjoyed updating this book and hope that it will be of some help as you continue to develop your HACCP and food safety management programme.

If you have enjoyed this book and want to give us feedback or want to pass on some suggestions on how it could be improved, then please contact us via the publishers. Or you can e-mail us at:

haccp.bluebook@btinternet.com

References, further reading and resource material

References

Abbott, H. (1992) *Product Risk Management*, Pitman Publishing, London.

AgriCanada (1994) Food Safety Enhancement Program, Implementation Manual Vol. 3: *Application of Generic HACCP Models*, AgriCanada, http://www.cfia-acia.agr.ca/english/food/haccp/haccp.html.

Beckers, H. J., Daniels-Bosman, M. S. M., Ament, A. *et al.* (1985) Two outbreaks of salmonellosis caused by *Salmonella indiana*. A survey of the European summit outbreak and its implications. *International Journal of Food Microbiology*, **2**, 185–95.

Bernard, D. T. and Scott, V. N. (1995) Risk assessment and food-borne microorganisms: the difficulties of biological diversity. *Food Control*, **6**, (6), 329–33.

Bird, M. (1992) *Effective Leadership*, BBC Business Matters Publications, London.

British Standards Institute (1995) *General Criteria for the Operation of Various Types of Bodies Performing Inspection*, BS EN 45004.

Brown, M. (1992) *Successful Project Management*, British Institute of Management, Hodder and Stoughton.

Calhoun County Department of Public Health *et al.* (1997) Hepatitis A associated with consumption of frozen strawberries – Michigan, March 1997. *Morbidity Mortality Weekly Rep.*, **46** (13), 288, 295.

Campden and Chorleywood Food Research Association (1992) *HACCP: A Practical Guide*, Technical Manual No. 38.

Campden and Chorleywood Food Research Association (1995a) *Considerations of Due Diligence with Respect to Pesticide Residues in Food and Drink*, Guideline No. 2, February 1995.

Campden and Chorleywood Food Research Association (1995b) *Guidelines for the Prevention and Control of Foreign Bodies in Food*, Guideline No. 5, August 1995.

Campden and Chorleywood Food Research Association (1996) *Assured Crop Production*, Guideline No. 10, April 1996.

Campden and Chorleywood Food Research Association (1997) *HACCP: A Practical Guide*, 2nd edn, Technical Manual No. 38.

CDR Weekly, Listeriosis Update (13), 26 March 1992.

Checkmate International (1997) Assured Produce Scheme, Checkmate International, Long Hanborough, Oxon, OX8 8LH, UK.

Chesworth, N. (ed.) (1997) *Food Hygiene Auditing*, Blackie Academic and Professional, London.

Codex Committee on Food Hygiene (1993) Guidelines for the Application of the Hazard Analysis Critical Control Point (HACCP) System, in *Training Considerations for the Application of the HACCP System to Food Processing and Manufacturing*, WHO/FNU/FOS/93.3 II, World Health Organization, Geneva.

Codex Committee on Food Hygiene (1997) *HACCP System and Guidelines for its Application*, Annex to CAC/RCP 1-1969, Rev. 3, in Codex Alimentarius Food Hygiene Basic Texts, Food and Agriculture Organization of the United Nations World Health Organization, Rome.

EURACHEM/WELAC (1993) *Accreditation for Chemical Laboratories: Guidance on the Interpretation of the EN 45000 Series of Standards and ISO/IEC, Guide 25*. (Available in the UK from Laboratory of the Government Chemist, Teddington, Middlesex, TW11 0LY.)

EEC Council Directive 89/107/EEC (1989) Concerning Food Additives Authorised for Use in Foodstuffs intended for Human Consumption.

European Community (1993) Council Directive 93/43/EEC (14 June) on the Hygiene of Foodstuffs, *Official Journal of the European Communities*, July 19, 1993, No. L 175/I.

European Community Council Directive 95/2/EC (1995) on food additives other than colours and sweeteners. *Official Journal of the European Communities*, 14 October 1995, No. L248.

Federal Register (1995) Procedures for the safe and sanitary processing and importing of fish and fishery products; final rule. *Federal Register*, **60**, (242), 65095–202.

Federal Register (1996) Pathogen reduction: hazard analysis and Critical Control Point (HACCP) Systems; final rule. *Federal Register*, **61**, (144), 38805–55.

Griffiths, C. J. and Worsfold, D. (1994) *Application of HACCP to Food Preparation Practices in Domestic Kitchens, Food Control*, Vol. 5, No 3, Butterworth–Heinemann, Oxford.

The Grocer (1991) 8 June. William Reed Publishing Ltd., Crawley, Sussex, UK.

Hathaway, S. (1995) *Harmonization of International Requirements under HACCP-based Food Control Systems, Food Control*, Vol. 6, No. 5, Butterworth–Heinemann, Oxford.

Haycock, P. and Wallin, P. (1998) Foreign Body Prevention, Detection and Control: A practical approach, Blackie Academic and Professional, London.

Hayes, G. D., Scallan, A. J. and Wong, J. H. F. (1997) Applying statistical process control to monitor and evaluate the hazard analysis critical control point hygiene data. *Food Control*, **8**, (4).

Hennessy, T. W., Hedber, C. W., Slutsker, L. *et al.* (1996) A national outbreak of *Salmonella enteritidis* infections from ice-cream. *New England Journal of Medicine*, **334**, (20),

Herwaldt, B. L. and Ackers, M. L., and Cyclospora Working Group (1997) An outbreak in 1996 of cyclosporiasis associated with imported raspberries. *New England Journal of Medicine*, **336**, 1548–56.

HMSO (1990) *Food Safety Act*, HMSO, London, UK.

HMSO (1993) *Assured Self Catering: A Management System for Hazard Analysis*, Department of Health, HMSO, London, UK.

HMSO (1997) *Food Safety Act 1990, Code of Practice No. 9: Food Hygiene Inspections*, HMSO, London, UK.

Hobbs, B. C. and Roberts, D. (1993) *Food Poisoning and Food Hygiene*, 6th edn, Edward Arnold, a Division of Hodder and Stoughton Ltd, Kent, UK.

Hourihane, J., Dean, T. and Warner, J. (1995) Peanut allergy: an overview. *Chemistry and Industry*, 299–302, UK.

ICMSF (1980) *Microbial Ecology of Foods*, Vol. 1, *Factors Affecting Life and Death of Micro-organisms*; Vol. 2, *Food Commodities*, Academic Press, New York.

ICMSF (1986) *Micro-organisms in Foods 2. Sampling for Microbiological Analysis: Principles and Specific Applications*, 2nd edn, Blackwell Scientific Publications, Oxford.

ICMSF (1988) *Micro-organisms in Foods 4. Application of the Hazard Analysis Critical Control Point (HACCP) System to Ensure Microbiological Safety and Quality*, Blackwell Scientific Publications, Oxford.

ICMSF (1996) *Micro-organisms in Foods 5. Microbiological Specifications of Food Pathogens*, Blackie Academic and Professional, London.

ICMSF (1998) *Microbial Ecology of Food Communities 6*. Blackie Academic and Professional, London.

IFST (1991) *Food and Drink – Good Manufacturing Practice: A Guide to its Responsible Management*, 3rd edn, Institute of Food Science and Technology, London.

IFST (1993) *Shelf Life of Foods – Guidelines for its Determination and Prediction*, Institute of Food Science and Technology, London.

International Standards Organization (1994a) ISO 9000 Series, *Quality Management and Quality Assurance Standards: Guidelines for Selection and Use.*

International Standards Organization (1994b) ISO 9001, *Quality Systems – Model for Quality Assurance in Design/Development, Production, Installation, and Servicing.*

International Standards Organization (1994c) ISO 9002, *Quality Systems – Model for Quality Assurance in Production and Installation.*

International Standards Organization (1994d) ISO 9003, *Quality Systems – Model for Quality Assurance in Final Inspection and Test.*

Jay, A. (1993) *Effective Presentation*, Pitman Publishing in association with the Institute of Management Foundation, London.

James, P. (1997) Food Standards Agency, An interim proposal, Cabinet Office, HM Government, UK.

Janofsky, M. (1997) 25 Million pounds of beef is recalled, New York Times, 22 August.

Kuritsky, J. N. *et al.* (1984) A community outbreak associated with bakery product consumption. *Annals of Internal Medicine*, **100**, 519–21.

Lee and Hathaway (1998) *The Challenge of Designing Valid HACCP Plans for Raw Food Commodities, Food Control*, Elsevier, Oxford, in press.

Lehmacher *et al.* (1995) *Epidemiology and Infection*, **115**, Cambridge University Press, Cambridge, UK.

Lloyds Register Quality Assurance Ltd. (1995) *Quality Systems for the Food and Drink Industries, Guidelines for the use of ISO 9001: 1994 in the Design and Manufacture of Food and Drink*, Issue 1.

Majewski, C. (1997) *Food Safety Strategies in the Changing Global Environment*, Study Fellowship 1996/97, European Commission, Brussels.

Mortimore, S. E. (1995) How effective are the current sources of food hygiene education and training in shaping behaviour? M.Sc. thesis, Leicester University, UK.

Mortimore, S. E. and Smith, R. A. (1998) Standardized HACCP training: assurance for food authorities. *Food Control*, Elsevier in press.

Mortimore, S. E. and Wallace, C. A. (1997) *A Practical Approach to HACCP, Training Programme*, Blackie Academic and Professional, London.

Mortimore, S. E. and Wallace, C. A. (1994) *HACCP: A Practical Approach*, Chapman & Hall, London.

Motarjemi, Y., Käferstein, F., Moy, G. *et al.* (1996) *Importance of HACCP for Public Health and Development, The Role of the World Health Organization, Food Control*, Vol. 7, No. 2, Elsevier, Oxford, pp. 77–85.

Murphy, J. (1997) Probe into dioxin-tainted animal feed results in alerts to producers. *Food Chemical News*, 14 July 1997, Washington, DC.

National Advisory Committee on Microbiological Criteria for Foods (1992) Hazard Analysis and Critical Control Point System (adopted 20 March 1992), *International Journal of Food Micro-biology*, **16**, 1–23.

National Advisory Committee on Microbiological Criteria for Foods (1997) *Hazard Analysis and Critical Control Point Principles and Application Guidelines* (adopted 14 August 1997).

North, R. and Gorman, T. (1998) An independent analysis of the salmonella in eggs scare, Health Unit Paper No. 10, IEA Health and Welfare Unit, London.

Notermans *et al.* (1995) The HACCP concept: specification of criteria using quantitative risk assessmnet. *Food Microbiology*, **12**, 81–90.

NSF International (1996) HACCP 9000, Worldwide Web:

http://www.nsf.com

Oates, D. (1993) *Leadership – The Art of Delegation*, The Sunday Times Business Series, Century Business, London.

The Pennington Group (1997) *Report on the Circumstances Leading to the 1996 Outbreak of Infection with E. coli O157 in Central Scotland, the Implications for Food Safety and the Lessons to be Learned*, HMSO, London, April 1997.

Pepsi-Cola Public Affairs (1993) *The Pepsi Hoax: What Went Right?* Pepsi-Cola, USA.

Pierson, M. D. and Corlett, D. A. (1992) *HACCP Principles and Application*, Van Nostrand Reinhold, New York.

Price, F. (1984) *Right First Time – Using Quality Control for Profit*, Gower, Aldershot.

Reuter (1990) France: Perrier puts cost of benzene scare at $79 million. Reuter Newswire – Western Europe, May 10.

Rogers, L. (1994) Salmonella Sandwich eaters win £200,000. *The Sunday Times*, 18 September, London.

Rowntree, D. (1981) *Statistics Without Tears*, Penguin, London.

Royal Institute of Public Health and Hygiene (RIPHH) (1995a) *The First Certificate in Food Safety*, RIPHH, 28 Portland Place, London.

Royal Institute of Public Health and Hygiene (1995b) *HACCP Principles and their Application in Food Safety* (IntroductoryLevel), Training Standard, RIPHH, 28 Portland Place, London.

Shapton, D. A. and Shapton, N. F. (1991) *Principles and Practices for the Safe Processing of Foods*, H. J. Heinz Company, Butterworth–Heinemann, Oxford.

Shapton, N. F. (1989) *Food Safety – A Manufacturer's Perspective*, Hobsons Publishing plc, Cambridge.

Socket, P. N. (1991) A review: The economic implications of human *Salmonella* infection. *Journal of Applied Bacteriology*, **71**, 289–95.

Sperber, W. H. (1995) Introduction to the use of Risk Assessment in Food Processing Operations: Background and Relationship to HACCP. Talk given at the IFT Annual Meeting Program, June, 1995.

Sprenger, R. (1995) *Hygiene for Management*, 7th edn, Highfield Publications, Rotherham and London.

USDA, Food Safety and Inspection Service (1996a) *Food Safety in the Kitchen: a 'HACCP' Approach*, USDA Consumer Publications, Washington, DC.

USDA, Food Safety and Inspection Service (1996b) 1. *Generic HACCP Model for Beef Slaughter*; 2. *Generic HACCP Model for Fully Cooked, Not Shelf Stable*, USDA, Washington, DC.

WHO (1992) *Toxic oil syndrome, current knowledge and future perspectives*, WHO, Geneva.

WHO (1995) *Training aspects of the Hazard Analysis Critical Control Point System (HACCP)*, WHO/FNU/FOS/96.3, WHO, Geneva.

Further reading

Bryan, F. (1981) Hazard Analysis of food service operations. *Food Technology*, **35**, (2), 78–87.

Doeg, C. (1995) *Crisis Management in the Food and Drink Industry: A Practical Approach*, Chapman & Hall, London.

Food Control, **5**, (3) Special issue of *Food Control* on HACCP, July 1994, Butterworth–Heinemann.

Food Quality and Standards Service, Food and Nutrition Division (1998) *Food Quality and Safety Systems*, A training manual on food hygiene, and the Hazard Analysis and Critical Control Point (HACCP) system. Food and Agriculture Organization of the United Nations, Rome.

IAMFES (1991) *Procedures to Implement the Hazard Analysis Critical Control Point System*, International Association of Milk, Food and Environmental Sanitarians Inc., Ames, Iowa, USA.

IFST (1990) *Guidelines for the Handling of Chilled Foods*, Institute of Food Science and Technology, London.

IFST (1992) *Guidelines to Good Catering Practice*, Institute of Food Science and Technology, London.

ILSI Europe (1997) *A Simple Guide to Understanding and Applying the HACCP Concept*, 2nd edn, International Life Sciences Institute Europe.

Mayes, T. (1992) Simple user's guide to Hazard Analysis Critical Control Point concept for the control of microbiological safety. *Food Control*, **3**, (1), 14–19.

Murphy, J. (1997) Dioxin investigation points to tainted clay added to chicken feed. *Food Chemical News*, **39** (19), 38.

Oakland, J. S. (1995) *Total Quality Management: Text with Cases*, Butterworth–Heinemann, Oxford.

Wilson, S. and Weir, G. (1995) *Food and Drink Laboratory Accreditation: A Practical Approach*, Chapman & Hall, London.

Computer software

Application of HACCP – computer program, Institute of Hygiene, Public Health Services and Management, Str. Dr Leonte 1–3, 76256 Bucharest, Romania.

CCFRA HACCP Documentation Software, Campden and Chorleywood Food Research Association, Chipping Campden, Gloucestershire GL55 6LD, UK.

doHACCP/recordHACCP, Celsis Ltd, Cambridge Science Park, Milton Road, Cambridge CB4 4FX, UK.

FIST–HACCP for Windows. A software package for the management of HACCP system. TNO Nutrition and Food Research Institute. BGT Division, PO Box 360, 3700 AJ Zeist, The Netherlands.

Food MicroModel, Leatherhead Food RA, Randalls Road, Leatherhead, Surrey KT22 7RY, UK.

RAMAS, Risk Assessment Management Application Software, Quality Systems Associates Ltd, PO Box 306, St Albans, Herts AL1 1YL, UK.

Databases

Dialog Information Service, DIALOG, PO Box 188, Oxford, OX1 5AX, UK.

Foodline, Leatherhead Food RA, Leatherhead, Surrey KT22 7RY, UK.

Food Science and Technology Abstracts, International Food Information Service (IFIS) Publishing, Lane End House, Shinfield, Reading, UK.

IPCS Intox System, World Health Organization, Geneva, Switzerland.

Joint FAO/WHO Codex Alimentarius Database, Maximum Residue Limits (MRLs), Chief, Joint FAO/WHO Food Standards Programme, FAO, Rome, Italy.

Medline, US National Library of Medicine, Library Operations, 8600 Rockville Pike, Bethesda, Maryland 20894, USA.

Reference Databases for Hazard Identification, Agriculture and Agri-Food, 59 Camelot Drive, Nepean, Ontario, Canada K1A 0Y9.

USFDA Prime connection, Center for Food Safety and Applied Nutrition, Food and Drug Administration, Washington, DC 20204, USA.

Videos

Critical Steps to Food Safety, Hazard Analysis and Control for Caterers, Retailers and Small Food Manufacturers, Shield Video Training Ltd, 159 Whiteladies Road, Bristol, BS8 2RF, UK.

Food Safe, Food Smart: HACCP and its Application to Food Industry, Agriculture, Food and Rural Development, Food Quality Branch, Edmonton, Alberta, Canada, T6H 4P2.

Food Safety: An Educational Video for Institutional Food Service Workers, Center for Food Safety and Applied Nutrition, Food and Drug Administration, Washington, DC, 20204, USA.

HACCP – A Retail Seafood Safety Program, New England Fisheries Development Association, Inc., 309 World Trade Center, Boston, Mass. 02210-2001, USA.

HACCP – How the System Works (1996), Reading Scientific Services Ltd./Shield Video Training Ltd., Reading Scientific Services Limited, The Lord Zuckerman Research Centre, Whiteknights, PO Box 234, Reading, RG6 6LA, UK. Tel: +44 (0) 118 986 8541; Fax: +44 (0) 118 986 8932.

HACCP video film, Agriculture and Agri-Food, 59 Camelot Drive, Nepean, Ontario, Canada, K1A 0Y9.

Links to food safety and HACCP Internet and WorldWide Web sites of interest

This review represents a selection of the available on-line information sources at a particular time. The structure of the electronic information super-highway is constantly evolving and there is no guarantee that these links will remain viable throughout the lifetime of this edition of the book. It is likely that anyone searching for information will choose a particular point of entry to the 'Web' from this list and will then find that most worthwhile sites will contain hypertext links which will transport them on to other relevant sites of interest, and thus will begin the journey of discovery. A few cautionary words are worthwhile. Many of the sites are accredited academic, governmental or other establishments, but not all. This is not to say that private sites or open mailing lists are any the less useful, but it is possible to publish opinion and 'research' which has not been subject to scrutiny or independently refereed. Thankfully there is not much of this around in the field of food safety, but as with everything else – user beware! Reference in this list to any specific commercial products, process, service, manufacturer or company does not constitute its endorsement or recommendation by the authors, who are not responsible for the contents of any pages referenced from this list. If you have any changes to this list to report, or have recommendations for sites to include in further editions, then please e-mail them to the authors at: haccp.bluebook@btinternet.com.

WorldWide Web sites

UK organizations

http://www.easynet.co.uk/ifst/ifsthp3.htm
This is the site of the UK Institute of Food Science and Technology. It provides useful opinion on a number of current issues (Hot Topics) and links to other food-related Web sites.

http://www.lfra.co.uk/index.html
Leatherhead Food Research Association – a membership-based and independent UK food research organization.

http://www.campden.co.uk/ccfra.htm
Campden and Chorleywood Food Research Association (CCFRA) – a membership-based and independent UK food research organization.

http://www.riphh.org.uk
You can find more about the Royal Institute of Public Health and Hygiene (RIPHH). This includes Hot Topics and details about hygiene training.

http://www.socgenmicrobiol.org.uk/
The Society for General Microbiology.

http://www.sofht.co.uk
The Society for Food Hygiene Technology.

USA organizations

http://www.ift.org/
Institute of Food Technologists in the USA. A non-profit scientific society with 28 000 members working in food science, food technology and related professions in industry, academia and government.

http://www.iit.edu/~ncfs/
National Center for Food Safety and Technology (NCFST) in the USA. NCFST is a consortium where academia, industry and government pool their resources to address the complex issues raised by emerging food technologies.

http://www.asmusa.org
American Society for Microbiology.

http://hammock.ifas.ufl.edu
Florida Agricultural Information Retrieval System.

http://ificinfo.health.org
International Food Information Council On-Line, and within this site:

- http://ificinfo.health.org/infosn.htm
  – IFIC's Food Safety and Nutrition Information page.

http://www.foodallergy.org/index.html
The Food Allergy Network (FAN) is a non-profit member-supported organization established to help families living with food allergies, and to increase public awareness about food allergies and anaphylaxis. FAN provides education, emotional support and coping strategies.

Rest of the world

http://www.dple.gov.au/ocvo.ofs.html
The Australian Office of Food Safety.

http://www.dfst.csiro.au/
The Division of Food Science and Technology, which is part of the
  Commonwealth Scientific and Industrial Research Organization
  (CSIRO) Food Processing Sector in Australia. During 1998, CSIRO is
  forming a joint venture with the Australian Food Industry Science
  Centre to form Food Science Australia. New web links will be
  announced. Within this site:

- http://www.dfst.csiro.au/programs/foodsafe.htm
  – CSIRO Food Safety and Hygiene Program. The work of the Program
  ranges from long-term strategic research projects to very short-term,
  problem-oriented investigations for individual companies.

http://fqp.al/haccp.pagina/
A HACCP information resource site in Dutch from independent consultants.

http://www.who.ch/
The World Health Organization in Geneva, Switzerland, and within this
  site:

- http://www.who.ch/programmes/fsf/index.htm
  – WHO Programme of Food Safety and Food Aid. Publications and
  documents on food safety. Links to other sites on food safety.
- http://www.fao.org/waicent/faoinfo/economic/esn/codex/
  codex.htm
  – on this page you will find information about the Codex Alimentarius and
  the Joint FAO/WHO Food Standards Programme. You will also find information about current events and forthcoming Codex meetings (including
  working papers) and summary reports of recently held Codex meetings.

http://www.nih.go.jp/yoken/index.html
The National Institute of Infectious Diseases in Japan.

http://foodnet.fic.ca/welcome.html
FoodNet, an information network for the food industry, is a joint pilot project
  of the Food Institute of Canada and Agriculture and Agri-Food Canada.

http://www.cfia-acia.agr.ca/
The Canadian Food Inspection Agency (CFIA). The government of Canada
  has consolidated all federally mandated food inspection and quarantine
  services into a single federal food inspection agency. Within this site:

- http://www.cfia-acia.agr.ca/english/food/haccp/haccp.html
  – the Food Safety Enhancement Program (FSEP) is an aid to Agriculture
  and Agri-Food Canada's implementation teams, its inspection workforce, and to industry's management and employees. Its use is intended
  during the implementation phases of HACCP. Includes HACCP checklist and generic HACCP models.

Government

http://www.maff.gov.uk/maffhome.htm
The UK Ministry of Agriculture, Fisheries and Food, covering farming,
  environment, fisheries, food, and public and animal health.

299

http://www.open.gov.uk/doh/dhhome/htm
The UK Department of Health.

http://www.fda.gov/
The Food and Drug Administration of the United States (FDA) – a
consumer protection agency.

http://vm.cfsan.fda.gov
FDA's Center for Food Safety and Applied Nutrition, and within this site:

- http://vm.cfsan.fda.gov/list.html
  – FDA's Center for Food Safety and Applied Nutrition, containing the
  National Food Safety Initiative and details of *Foodborne Illness and the
  Bad Bug Book* (foodborne pathogenic micro-organisms and natural
  toxins).
- http://vm.cfsan.fda.gov/~mow/intro.html
  – *The Bad Bug Book*. This handbook provides basic facts regarding food-
  borne pathogenic microorganisms and natural toxins.

http://www.cvm.fda.gov
FDA Center for Veterinary Medicine, includes a searchable database for
FDA-approved animal drug products.

http://www.nal.usda.gov/fnic/index.html
  – The Food and Nutrition Information Center (FNIC) is one of several
  information centres at the National Agricultural Library (NAL),
  Agricultural Research Service (ARS), United States Department of
  Agriculture (USDA). And within this site:
- http://www.nal.usda.gov/fnic/foodborne/fbindex/index.htm
  – source directory on foodborne illness information;
- USDA/FDA HACCP Training Programs and Resources Database.

http://cos.gdb.org/best/fedfund/usda/usda-intro.html
USDA Research Database.

http://www.cdc.gov
Center for Disease Control, and within this site:

- http://www.cdc.gov/ncidod/ncid.htm
  – National Council for Infectious Diseases, information on the causes
  and prevention of infections diseases, including foodborne diseases.
- http://www.cdc.gov/ncidod/diseases/foodborn/foodborn.htm
  – the NCID conducts surveillance, epidemic infestigations, epidemio-
  logic and laboratory research, training and public education
  programmes to develop, evaluate and promote prevention and control
  strategies for infectious diseases.
- http://www.cdc.gov/ncidod/dbmd/foodborn.htm
  – National Center for Infectious Diseases, Centers for Disease Control
  and Prevention, Atlanta, GA, information and fact sheets on foodborne
  and diarrhoeal diseases.

http://www.usda.gov/
The United States Department of Agriculture, and within this site:

- http://www.usda.gov/mission/fs.htm
  – USDA's Food Safety organization, which places increased emphasis on pathogen reduction and hazard analysis and on critical control points (HACCP) in the entire meat and poultry production chain.
- http://www.usda.gov/fsis/
  – United States Department of Agriculture Food Safety and Inspection Service (FSIS), a public health agency in the US Department of Agriculture, which seeks to protect consumers by ensuring that meat and poultry products are safe, wholesome and accurately labelled.
- http://www.usda.gov/fsis/imphaccp.htm
  – general information about HACCP regulation and implementation in the USA.

http://www.econ.ag.gov/prodsrvs/rept-fd.htm
Information on the economics of food safety, including the estimated annual costs of infections, and other economic issues associated with food safety.

http://www.epa.gov/internet/oppts
Environmental Protection Agency, Office of Prevention, Pesticides and Toxic Substance, information about pesticides: regulation, use, toxicity, health risks.

http://www.aphis.usda.gov
USDA's Animal and Plant Health Inspection Service, information on pathogens and diseases at the farm level; information on regulation and inspection in the meat and poultry industry.

Academic

http://www.foodsafety.org/index.htm
The USA National Food Safety Database at the University of Florida, funded by the USDA. The goal of the National Food Safety Database project is to develop an efficient management system of US food safety databases that are used by consumers, industry and other public health organizations.

http://www.uark.edu/depts/fsc/
The Food Safety Consortium consists of researchers from the University of Arkansas, Iowa State University and Kansas State University. Conducts extensive investigation into all areas of poultry, beef and pork meat production from the farm to the table.

http://foodsafe.ucdavis.edu/homepage.html
The food safety site of the University of California at Davis with Hot Topics on food safety and links to food safety related sites and an Experts Directory:

- http://www.seafood.ucdavis.edu/haccp/plans.htm
  – Seafood Network Information Center including generic HACCP Plans.
- http://www.seafood.ucdavis.edu/haccp/ha.htm
  – National Seafood HACCP Alliance for Training and Education in the

USA, which was initiated by the Association of Food and Drug Officials (AFDO) and their regional affiliate of Southern States (AFDOSS) in conjunction with a cadre of sea Grant Seafood Specialists which originally assisted the National Fisheries Institute (NFI) with their initial HACCP training programmes. This is the source of the Seafood HACCP mailing list (see below).

http://ifse.tamu.edu/haccpall.html
The Center for Food Safety at Texas A&M University. The International Meat and Poultry HACCP Alliance is within the Institute of Food Science and Engineering's (IFSE) Center for Food Safety. The International Meat and Poultry HACCP Alliance was developed to provide a uniform programme to assure safer meat and poultry products.
- http://ifse.tamu.edu/alliance/haccpmodels.html
  – contains generic HACCP models and guidebooks from FSIS which are all saved as Adobe™ PDF files. You can download an Adobe PDF Reader Internet plug-in directly from Adobe Systems.
- http://bbq.tamu.edu/hrds/ssop.html
  – downloadable Standard Operating Procedures for Sanitation (Sanitation SOPs) in Federally Inspected Meat and Poultry Establishments. The Food Safety Inspection Service (FSIS) is pursuing a broad and long-term science-based strategy to improve the safety of meat and poultry products to better protect public health. They require meat and poultry establishments to develop and implement a written Standard Operating Procedure for sanitation (Sanitation SOPs) which addresses these areas.

Mailing lists

As well as Worldwide Web sites, which may be viewed at any time, there are also a number of mailing lists to which individuals may subscribe by e-mail. These function either as discussion areas, where opinions on various current topics can be aired, or as an early warning system for the rapid dissemination of vital information, for example emerging foodborne illness outbreaks. Three lists are given here which are central to food safety and HACCP, but there are also lists available on related topics such as microbiological research and public health epidemiology. The user will rapidly become aquainted with these from cross-postings to multiple lists. Although contributions to these lists are welcome and encouraged, it is also permissible to join the lists and lurk in the background, learning from the information and views expressed.

Foodsafe

This is an unmoderated, open access, discussion list maintained by the USDA/FDA Foodborne Illness Education Information Center, ARS National Agricultural Library in co-operation with the USDA Food

Safety and Inspection Service (FSIS) and the Food and Drug Administration (FDA).

To subscribe to Foodsafe

Send an e-mail to: majordomo@nal.usda.gov
Leave the subject line blank
In the body of the message type:
    subscribe foodsafe <firstname lastname> <your e-mail address>
For example:
    subscribe foodsafe Janet Jones janet.jones@provider.com
Turn off any automatic signature

Seafood HACCP

Likewise, this is an unmoderated, open access, discussion list maintained by the University of California at Davis.

To subscribe to Seafood HACCP

Send an email to: listproc@ucdavis.edu
Leave the subject line blank
In the body of the message type: subscribe seafood <firstname lastname>
For example: subscribe seafood Janet Jones
Turn off any automatic signature

FSnet

This is a moderated list produced by researchers at the Science and Society Project at the University of Guelph in Canada. It abstracts mailings from other lists and cuttings from various press reports and compiles them into a bulletin which is issued several times each week.

To subscribe to FSnet

To subscribe to FSnet, send mail to: listserv@listserv.uoguelph.ca
Leave the subject line blank
In the body of the message type: subscribe fsnet-L <firstname lastname>
For example: subscribe fsnet-L Janet Jones
Turn off any automatic signature

This list requires resource to make it happen and is part of a fundamental research program in the use of electronic information as a risk analysis tool. Participation is free of charge but financial contributions are welcome. It is supported by a mix of private and public-sector funding for which recognition is given. You may contact: Dr Douglas Powell, Director, Science and Society Project, University of Guelph, Guelph, Ontario N1G 2W1, Canada;

Tel: 519-821-1799; Fax: 519-824-6631; e-mail: dpowell@uoguelph.ca

Appendix A   Case studies

A.1 Introduction

Five practical case studies have been constructed to illustrate the application of the HACCP Principles to different areas of food and drink production and preparation. The authors of these case-study examples are people within the food and drink industry, who have hands-on experience of implementing HACCP. Each example has been carefully chosen so that this appendix represents a wide range of process environments and technologies, and products have been included which are normally considered 'high' or 'low' risk from the product safety viewpoint.

- Case Study A.2 Fishball – medium-scale manufacturing, automated – high risk.
- Case Study A.3 Long Goods pasta – large-scale maufacturing, automated – medium risk.
- Case Study A.4 Beef slaughter and dressing – medium-scale, hands on – high risk.
- Case Study A.5 Community meals – small-scale, hands on – high risk.
- Case Study A.6 Chocolate Mini Eggs – large-scale, automated – low risk.

Note: each case study detailed here is theoretical and the findings may not be exhaustive. The contributors are experienced in the products concerned, but the case studies do not necessarily reflect the views of approaches taken by their companies, nor those of the book authors. The examples are not intended as specific recommendations for similar processes/products, i.e. they are not generic HACCP Plans, but as a demonstration of the application of the HACCP Principles in different situations.

Some points to consider when looking at the case studies are:

- Several different styles are represented. HACCP does not always have to be documented in the same regimented way, but can follow a company style.
- HACCP Teams do not always use the CCP Decision Tree, and may rely on the experience of team members. However, when it is used, the decision tree is a great help in structuring thinking and checking decisions.
- Scope of the HACCP Study must always be clearly defined.

This case study is an example to illustrate the HACCP approach and is provided without any liability whatsoever in its application and use.

A.2 Case study – fishball

Andrew Mak and K.Y. Chong, Singapore Productivity and Standards Board, PSB Building, 1 Science Park Drive, Singapore 118221

A.2.1 Introduction

A family-owned medium-sized seafood processing company has been asked by its new customer in Europe to provide a HACCP Plan for its main product – boiled fishball.

A.2.2 HACCP Team members

- Production Manager: 12 years of fishball manufacturing experience.
- Production Supervisor: 5 years of fishball manufacturing experience.
- Quality Assurance Manager: started working in the company 3 years ago, HACCP trained, selected as Team Leader.
- Assistant Quality Assurance Manager: 4 years of working experience, HACCP trained, to act as Technical Secretary of the team.

A.2.3 Terms of reference

The HACCP study covers all types of food safety hazards, namely biological, chemical and physical. The HACCP study did not include any cleaning operations as it was covered by the factory-wide Good Manufacturing Practices (GMP) procedures.

A.2.4 Product description

• General

Fishballs are produced and consumed widely in Asia. They could be made from fish fillets or *surimi* or a combination of both. Fishballs are usually boiled briefly in water/gravy or deep-fried before consumption. They could be eaten as snacks, or served as a dish, cooked with noodles, rice or vegetables.

Surimi is the mixture of a wide variety of minced fish meat with added binding agents (e.g. polyphosphates) and sugar. The types of fish used include threadfin bream, bigeye snapper, jewfish and

This case study is an example to illustrate the HACCP approach and is provided without any liability whatsoever in its application and use.

lizard fish. It is mainly produced in Thailand and exported as frozen semi-processed materials. It can be used in many kinds of minced fish products such as fishballs/fishcakes, breaded fishburger, cuttle-fish-balls, prawn-balls, imitation crab meat (or crab-sticks), etc.

• Ingredients

Frozen *surimi*, frozen fish paste, salt, sugar, monosodium glutamate, polyphosphates, potato starch, water.

• Process

The processing of fishballs involves mixing of ingredients in a bowl chopper (or mixing machine), shaping of fishballs using a forming machine, setting in a water bath, cooking in boiling water, cooling in an air-conditioned tunnel and individually quick-freezing in a freezing tunnel. The finished products are then heat-sealed in plastic bags, date-marked and kept in a freezer at –20 °C until despatch.

• Product specifications

Microbiological:

Total viable count (cfu/g):	<100 000
E. coli (cfu/g):	<100
S. aureus (cfu/g):	<100
V. parahaemolyticus (in 50 g):	not detected
Salmonella (in 50 g):	not detected
Shigella (in 50 g):	not detected

Pack size: 300 g per pack. Minimum 50 pieces.
Shelf-life: 1 year from the date of production.
Storage and distribution: below –20 °C.

• Intended use

General public. Consumer instructions are as follows:
Keep frozen until ready to use. Cook in boiling water or deep-fry for about 2 minutes until thoroughly cooked at the centre. Serve as a snack or together with rice, noodles or vegetables.

This case study is an example to illustrate the HACCP approach and is provided without any liability whatsoever in its application and use.

A.2.5 *Process Flow Diagram*

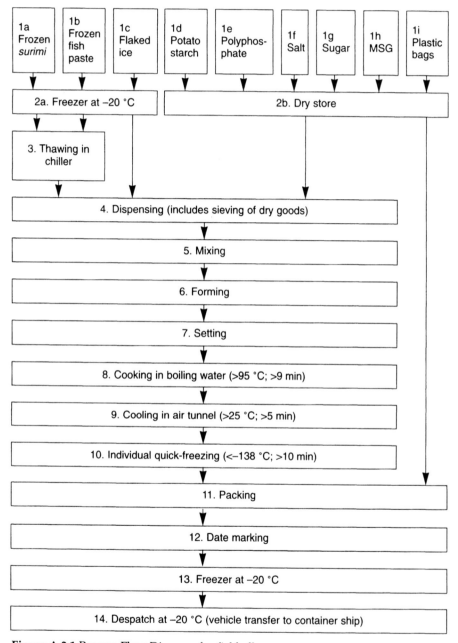

Figure A.2.1 Process Flow Diagram for fishballs.

A.2.6 Hazard Analysis

- **Microbiological (M)**

  (a) Infectious vegetative pathogens include *E. Coli, V. para-haemolyticus, Salmonella* and *Shigella*. Although a final cooking step at the consumer's end is assumed, the HACCP Team decided to include these organisms as a safety precaution.

  (b) The team concluded that the only toxin-producing pathogen of significance in this study is *S. aureus*.

  (c) Spoilage organisms such as *Pseudomonas, Acinetobacter* and *Moraxella* were not considered in the study as they are not pathogenic in nature.

- **Chemical (C)**

  (a) *Detergent and sanitizer residues.

  (b) *Lubricants from machinery, conveyors, etc.

- **Physical (P)**

  (a) Fish bones and scales from *surimi* and fish paste.

  (b) Metal fragments from machinery such as the mincer.

  (c) *Glass from lighting, equipment panels, etc.

  (d) *Wood chips from pallets, fittings, etc.

  (e) *Other foreign objects from personnel and environment, e.g. hair, jewellery, pen tops

  *These hazards have previously been considered by the HACCP Team and will be managed under the factory-wide Good Manufacturing Practice (GMP) procedures.

A.2.9 Implementation and maintenance

Since this is the company's first HACCP Plan, the HACCP Team decided to review the whole plan, including the controls of CCPs, after 1 month from implementation. By this time they hope to have approval to purchase a new metal detector for the packing room. This would mean a change to CCP 5 and a HACCP Plan amendment.

This case study is an example to illustrate the HACCP approach and is provided without any liability whatsoever in its application and use.

Regular audit will also be carried out every 3 months, and an annual audit by external third party. The audit will include a review of CCP log sheets, Certificates of Conformance from suppliers, maintenance records and quarantine records. The results of the audits will be discussed at the senior management meetings.

This case study is an example to illustrate the HACCP approach and is provided without any liability whatsoever in its application and use.

A.2.7 *Identification of CCPs*
Table A.2.1 HACCP Control Chart: hazard analysis and CCP determination

| Company: Tasty Fishballs Ltd | Date: 1st October 97 |
| Product: Boiled fishballs (for demonstration only) | Authorized by: Mr Vip |

No.	Process step	Potential hazards and possible causes	Control measures	Decision tree (Codex, 1997)				CCP
				Q1	Q2	Q3	Q4	(Y/N)
1	Receiving of raw materials							
1a	Frozen *surimi*	M: Presence of pathogenic organisms: *Salmonella, Shigella, Vibrio parahaemolyticus, Escherichia coli*	Approved suppliers provide Certificate of Conformance by batch Effective cooking at step 8	Y	N	Y	Y (8)	N
		P: Presence of fish bones and scales	Approved suppliers provide Certificate of Conformance by batch Correct setting of bowl chopper at step 5	Y	N	Y	Y (5)	N
1b	Frozen fish paste	M: Presence of pathogenic organisms: *Salmonella, Shigella, Vibrio parahaemolyticus, Escherichia coli*	Approved suppliers provide Certificate of Conformance by batch Effective cooking at step 8	Y	N	Y	Y (8)	N
		P: Presence of fish bones and scales	Approved suppliers provide Certificate of Conformance by batch Correct setting of bowl chopper at step 5	Y	N	Y	Y (5)	N

No.	Process step	Potential hazards and possible causes	Control measures	Decision tree (Codex, 1997)				CCP
				Q1	Q2	Q3	Q4 (Y/N)	
1c	Flaked ice	M: Presence of coliforms	Use only mains water which has been chlorinated Effective cooking at step 8	Y	N	Y	Y (8)	N
1d	Potato starch	C & P: Presence of impurities	Approved suppliers provide Certificate of Conformance by batch Visual inspection at the point of receiving and sieving at dispensing	Y	N	Y	Y (4)	N
1e	Polyphosphate	C & P: Presence of impurities	Approved suppliers provide Certificate of Conformance by batch Visual inspection at the point of receiving and sieving at dispensing	Y	N	Y	Y (4)	N
1f	Salt	C & P: Presence of impurities	Approved suppliers provide Certificate of Conformance by batch Visual inspection at the point of receiving and sieving at dispensing	Y	N	Y	Y (4)	N

| Company: Tasty Fishballs Ltd | | | Date: 1st October 97 | | | | |
| Product: Boiled fishballs (for demonstration only) | | | Authorized by: Mr Vip | | | | |

| No. | Process step | Potential hazards and possible causes | Control measures | Decision tree (Codex, 1997) | | | | CCP |
				Q1	Q2	Q3	Q4	(Y/N)
1g	Sugar	C & P: Presence of impurities	Approved suppliers provide Certificate of Conformance by batch. Visual inspection at the point of receiving and sieving at dispensing	Y	N	Y	Y (4)	N
1h	MSG	C & P: Presence of impurities	Approved suppliers provide Certificate of Conformance by batch. Visual inspection at the point of receiving and sieving at dispensing	Y	N	Y	Y (4)	N
1i	Plastic bags	C & P: Presence of impurities	Approved suppliers provide Certificate of Conformance by batch. Visual inspection at the point of receiving and packing	Y	N	Y	Y (11)	N
2a	Freezer	M: Growth of pathogenic organisms	Correct setting of freezer temperature (< −20 °C). Observe good storage practices. First-in-first-out approach	Y	N	Y	Y (8)	N

No.	Process step	Potential hazards and possible causes	Control measures	Decision tree (Codex, 1997)				CCP
				Q1	Q2	Q3	Q4	(Y/N)
2b	Dry store	M: Growth of pathogenic organisms	Keep dry store well ventilated. Observe good storage practices. First-in-first-out approach	Y	N	Y	Y (8)	N
3	Thawing in chiller	M. Growth of pathogenic organisms	Correct setting of chiller temperature (< 10 °C)	Y	N	Y	Y (8)	N
4	Dispensing	M: Growth and contamination of pathogenic organisms	Observe food hygiene practices	Y	N	Y	Y (8)	N
		P: Presence and contamination of foreign bodies	Observe food hygiene practices.	Y	N	Y	N	Y-CCP

Table A.2.1 *continued*

Company: Tasty Fishballs Ltd		Date: 1st October 97
Product: Boiled fishballs (for demonstration only)		Authorized by: Mr Vip

No.	Process step	Potential hazards and possible causes	Control measures	Decision tree (Codex, 1997)				CCP
				Q1	Q2	Q3	Q4	(Y/N)
5	Mixing	M: Growth and contamination of pathogenic organisms	Oberve food hygiene practices. Effective cooking at step 8	Y	N	Y	Y (8)	N
		P: Metal fragments from chopping blades	Adherence to maintenance schedule. Check integrity of chopping blades regularly [HACCP team recommends metal detection after date-marking of products]	Y	N	Y	N	Y-CCP
6	Forming	M: Growth and contamination of pathogenic organisms	Oberve food hygiene practices. Effective cooking at step 8	Y	N	Y	Y (8)	N
7	Setting	M: Growth and contamination of pathogenic organisms	Strict adherence to setting temperature and time. Effective cooking at step 8	Y	N	Y	Y (8)	N
8	Cooking	M: Survival of pathogenic organisms	Strict adherence to scheduled cooking temperature and time	Y	Y	–	–	Y-CCP

No.	Process step	Potential hazards and possible causes	Control measures	Decision tree (Codex, 1997)				CCP
				Q1	Q2	Q3	Q4	(Y/N)
9	Cooling	M: Contamination through equipment and operators	Strict adherence to scheduled cooling temperature and time. Observe food hygiene practices. Avoid touching foods with bare hands and soiled utensils	Y	N	Y	N	Y-CCP
10	Individual quick freezing	M: Contamination through equipment and operators	Strict adherence to scheduled cooling temperature and time. Observe food hygiene practices. Avoid touching food with bare hands and soiled utensils	Y	N	Y	N	Y-CCP
11	Packing	M: Contamination through equipment and operators	Observe food hygiene practices. Avoid touching foods with bare hands and soiled utensils. Check integrity of seals before start-up and then regularly	Y	N	Y	N	Y-CCP
12	Date marking	M: Loss of traceability due to illegible coding	Proper maintenance of equipment. Visual inspection of production and best before date	–	–	–	–	Y-CCP

Table A.2.1 *continued*

Company: Tasty Fishballs Ltd		Date: 1st October 97		
Product: Boiled fishballs (for demonstration only)		Authorized by: Mr Vip		

No.	Process step	Potential hazards and possible causes	Control measures	Decision tree (Codex, 1997)				CCP
				Q1	Q2	Q3	Q4	(Y/N)
13	Freezer	M: Growth of pathogenic organisms	Correct setting of freezer temperature (< 20 °C)	Y	N	Y	N	Y-CCP
14	Distribution	M: Growth of pathogenic organisms	Correct setting of freezer temperature (< 20 °C)	Y	N	Y	N	Y-CCP

A.2.8 Controlling CCPs

Table A.2.2 HACCP Control Chart: controlling of CCPs

| | | ISSUE NO: 1 | PAGE: 1 OF 3 |

Company: Tasty Fishballs Ltd — Date: 1st October 97

Product: Boiled fishballs (for demonstration only) — Authorized by: Mr Vip

No.	Critical limits (What)	Monitoring procedures			Corrective actions	Documentation
		How	When	Who		
4	P: Absence of physical impurities	Visual inspection Check compliance with hygiene procedures	Each batch Weekly line audit	Operators Supervisor	Remove impurities and record incidence. Change supplier/re-train operators/quarantine affected batch as necessary	QC checklists training records. Quarantine records
5	P: Chopping blades intact and in good condition	Visual inspection	At start and end of production – each shift	Operators	Change chopping blades immediately Quarantine previous batch if necessary Report to QA/maintenance	Maintenance records. Quarantine records
8	Cooking water >95 °C; cooking time >9 min. Product centre temperature 79 °C	– 2 temperature probes fixed at the start and end of the conveyor. Conveyor speed fixed, double-checked by operators with digital thermometer and stop watch – Check with hand-held probe	– Continuous monitoring and recording of water temperature; core product temperature and conveyor speed checked twice a day – Each batch	Operators Operators	Increase temperature setting or/and extend cooking time by slowing down conveyor speed. Check core temperature of products manually, quarantine batch if necessary	QC checklists, thermographs of water temperature. Quarantine records

Table A.2.2 *continued*

Company: Tasty Fishballs Ltd	Date: 1st October 97
Product: Boiled fishballs (for demonstration only)	Authorized by: Mr Vip

No.	Critical limits (What)	Monitoring procedures			Corrective actions	Documentation
		How	When	Who		
9	Air temperature of cooling tunnel <25 °C; cooling time 5 min. Product temperature at exit <10 °C	– 2 temperature probes fixed at the start and end of the conveyor. Conveyor speed fixed, double-checked by operators with digital thermometer and stop watch – Check with hand-held probe	– Continuous monitoring and recording of air temperature; core product temperature and conveyor speed checked twice a day – Each batch	Operators Operators	Lower temperature setting or/and extend cooling time by slowing down conveyor speed. Check core temperature of products manually, quarantine batch if necessary	QC checklists, thermographs of air temperature in cooling tunnel. Quarantine records
10	Air temperature of IQF tunnel < –138 °C; cooling time > 10 min	2 temperature probes fixed at the start and end of the conveyor. Conveyor speed fixed, double-checked by operators with digital thermometer and stop watch	Continuous monitoring and recording of air temperature; core product temperature and conveyor speed checked twice a day	Operators	Lower temperature setting or/and extend cooling time by slowing down conveyor speed. Check core temperature of products manually, quarantine batch if necessary	QC checklists, thermographs of air temperature in cooling tunnel. Quarantine records

No.	Critical limits	Monitoring procedures			Corrective actions	Documentation
	(What)	How	When	Who		
11	No touching of finished products with bare hands	Visual inspection	Continuous	Line supervisors	Retrain operators. Dispose of contaminated products	QC checklists, training records
12	Clear and legible coding	Visual inspection	Before start-up and once per shift	Line supervisors	Hold product and re-date. Input correct code or repair accordingly	QC checklists Maintenance records
13	Freezer <–20 °C	Temperature probe fixed at the freezer. Operators double-checked by digital thermometer	Continuous monitoring and recording. Air temperature manually checked twice a day	Operators	Report to Production Manager immediately, call for repair and move goods to alternate freezer of necessary	Thermograph of freezer. QC checklists Maintenance records Quarantine records
14	Distribution <–20 °C	Temperature probe fixed in container	Continuous monitoring and recording	Drivers	Quarantine batch, report to Quality Assurance Manager immediately	Thermograph Quarantine records

A.3 Case study – Long Goods Pasta Product

Pablo de Jongh G., with collaboration of the Catia La Mar Food Safety Committee, Cargill de Venezuela, Catia La Mar Pasta Plant

A.3.1 Introduction

Cargill de Venezuela is a subsidiary of Cargill Inc., Minneapolis, Minnesota, USA. It began its operations 11 years ago, and operates wheat mills, rice mills, pasta plants and oil seed plants.

This case study describes a HACCP Plan for one of the dry milling plants, Catia La Mar Plant, which has a wheat mill and a pasta production plant. Specifically we will describe the HACCP Plan for long goods pasta, processed in a high-temperature pasta line.

A.3.2 HACCP Team members

At Cargill de Venezuela we have administered food safety by committees in each plant. All of these committees are co-ordinated by a Divisional Food Safety Coordinator who reports to the Corporate Food Safety Coordinator in Minneapolis. Each plant has a HACCP Team for specific production lines.

The Catia La Mar Food Safety committee is organized as follows:

- Plant Manager
- Pasta Manager
- Mill Manager
- Quality Control Supervisor
- Logistics Supervisor
- Quality Assurance Supervisor.

The HACCP Team for the pasta plant is organized as follows:

- Pasta Plant Production Manager
- Production Supervisors (high-temperature pasta line operators)
- Packing Supervisors
- Maintenance Supervisors.

This case study is an example to illustrate the HACCP approach and is provided without any liability whatsoever in its application and use.

A.3.3 Terms of reference

This case study covers all types of hazards: physical (foreign material), chemical and microbiological. Not included in this case study are the prerequisite programmes, or Good Manufacturing Practices that support our HACCP System. These are plant sanitation, personal hygiene, purchasing, water/ice, air, pest control, maintenance, labelling, transportation, rework and retrieval.

A.3.4 Product description

Long goods pasta is a dried product, not fresh pasta, with a maximum moisture content of 13%. It is made from a combination of one or more of the following raw materials: durum semolina, semolina (from hard wheat), durum flour, wheat flour, regrind pasta and water. All raw materials comes from our own mill, located next to the pasta plant. Long goods pasta or spaghetti (cord shaped, not tubular, and diameter between 0.13 and 0.18 m) is rigid when it is uncooked. Very important quality factors to be considered during this process are colour before cooking, which must be uniform yellow without black or white points; firmness to the bite and colour after cooking.

Dried raw materials are weighed and mixed. Then water and dried materials are mixed to form a dough. The dough goes through an extruder and former where the pasta is shaped. It then passes through the pre-dryer and dryer, where the moisture goes down by the action of temperature (80 °C) and time (6 hours). This product is cooled to ambient temperature before storage. The product is packaged in 1 kg and 0.5 kg polypropylene bags, and shrink wrapped in 6 or 12 kg polyethylene bags.

This product is intended to be for general public consumption.

Foreign materials are the main hazard encountered in our industry but we have not ignored the chemical and microbiological hazards in our hazard analysis. Raw materials coming from our own mill are one of the most important parts of the process. In order to assure that the incoming raw materials that go into the pasta plant are safe, HACCP Plans were also developed for the flour mill.

This case study is an example to illustrate the HACCP approach and is provided without any liability whatsoever in its application and use.

A.3.5 The process

The production process of the Long Goods Pasta Product is shown in the Process Flow Diagram (Figure A.3.1).

A.3.6 Hazard analysis

The hazard analysis was done from the Process Flow Diagram, and is shown in Table A.3.1.

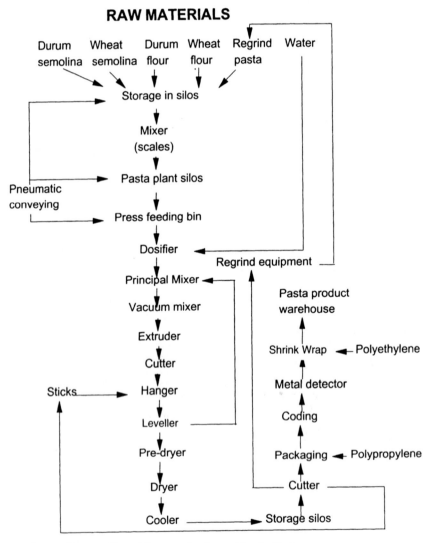

Figure A.3.1 Process Flow Diagram for Long Goods Pasta.

Table A.3.1 Hazard analysis

Process step	Hazard	Preventative measure
1 Raw materials		
Durum semolina	Foreign material	Durum Mill HACCP Plan
Semolina	Foreign material	Wheat Mill HACCP Plan
Wheat flour	Foreign material	Wheat Mill HACCP Plan
Durum flour	Foreign material	Durum Mill HACCP Plan
Reground pasta	Foreign material	Sifters screen inspection Visual inspection of product
Water	Microbiological Chemical Foreign materials	Water analysis of the incoming water Water tank cleaning in concordance with the GMP plan
2 Raw materials storage silos	Foreign materials	Silo cleaning in concordance with the GMP plan Monthly hygiene audit
3 Raw materials mixer	Foreign materials	Mixer cleaning in concordance with the GMP plan Monthly hygiene audit
4 Pasta plant silos	Foreign material	Silo cleaning in concordance to the GMP plan Monthly hygiene audit
5 Press feeding bin	No hazard identified	
6 Dosing	No hazard identified	
7 Principal mixer	Microbiological	Daily cleaning by scraping with spatula
8 Vacuum mixer	Microbiological	Daily cleaning by scraping with spatula
9 Extrusion	No hazard identified	
10 Former	No hazard identified	
11 Cutter	No hazard identified	

Table A.3.1 *continued*

Process step	Hazard	Preventative measure
12 Hanger	No hazard identified	
13 Leveller	No hazard identified	
14 Pre-drying	No hazard identified	
15 Drying	No hazard identified	
16 Cooling	No hazard identified	
17 Storage	Foreign material Chemical	Cleaning and fumigation in concordance with the GMP plan Fumigation without pasta into the storage silo and enough aeration time
18 Cutter	No hazard identified	
19 Packaging	No hazard identified	
19.1 Polypropylene	Foreign material Chemical	Certificate of Analysis from suppliers
20 Metal detector	Metal contamination not detected	Periodical check to confirm adequate equipment function
21 Coding	Product without code	Periodical check to confirm adequate equipment function
22 Shrink wrap machine	No hazard identified	
22.1 Polyethylene	No hazard identified	
23 Pneumatic conveying (air)	Foreign material	Change air filter periodically

A.3.7 Critical Control Point identification

Critical Control Points were identified by the HACCP Team using the CCP Decision Tree. The CCPs identified by the team were revised by Catia La Mar Plant 'Food Safety Committee'. The CCPs are shown in Table A.3.2.

This case study is an example to illustrate the HACCP approach and is provided without any liability whatsoever in its application and use.

Table A.3.2 HACCP Plan for Long Goods Pasta: CCPs description

Process step	CCP No.	Hazard	Monitoring			Critical limits	Corrective actions	Verification			Action	Justification	
			Action	Responsible	Frequency			Action	Responsible	Frequency		Responsible	Frequency
Principal and vacuum mixers	1	Micro-biological	Scrape with spatula	Pasta line operator	Every day	Mixer walls free of moulds	Stop line, clean and sanitize it	Check the mixers	Production Supervisor	Once per shift	Micro. Analysis	Quality Control lab.	Once per month
Coding	2	Illegibile or inaccurate	Random check	Packaging operator	5 samples per hour	Production code and price in concordance with the product	Reject packages, advise supervisor, shut down line and review previous production	Random check	Quality Control Inspector	Twice per shift	Audit in Warehouse	Quality Control Manager	Monthly
Metal detector	3	Foreign material	Test detector with ferrous metal sample card of 1.2 mm Record in log sheet	Packaging machine operator	Once each shift	Metal detector fully functional, detects 1.2 mm ferrous sphere	Notify supervisor immediately, re-check all rejected packages	Check metal detector log for correct entries	Shift Supervisor	Once each day	No complaints	Plant Manager	Monthly

A.3.8 CCP Management

The CCPs identified for long goods pasta production are described in Table A.3.2, indicating who is responsible for monitoring procedures and frequency, the critical limits and corrective actions to be taken in case of a deviation, and who is responsible for the verification and its frequency. At Cargill we have an extra step in our HACCP Plans, the justification, to assure the good management of the HACCP Plan. During development of each CCP responsibility for the justification and its frequency is included.

Log sheets were developed for each CCP. The sheets are located at the monitoring point. The CCP log sheet also describes where the completed sheets are to be filed.

Responsibility for monitoring, verification and justification is included in our employee HACCP training programme.

A.3.9 HACCP Plan maintenance

The justification process helps manage our HACCP Plan maintenance.

Food safety audits done by the Food Safety Coordinator, and the Monthly Hygiene Audits that include a review of the CCPs, allow us to determine when the HACCP Plan needs to be revised. Periodic revisions are made by the Plant Food Safety Committee.

This case study is an example to illustrate the HACCP approach and is provided without any liability whatsoever in its application and use.

A.4 Case study – beef slaughter and dressing

Alison Sloan, Foyle Meats Ltd, Lisahally, Campsie, Londonderry, BT47 1TJ, Northern Ireland

A.4.1 Statement of policy

Our policy is to provide an extremely high standard of service to all our customers and to ensure that high standards of food safety are maintained at all times.

In order to achieve our aims this HACCP Plan has been produced from a full hazard analysis of our operations and is designed to control any risk to food safety.

We are fully committed to the implementation and maintenance of this system and have put in place strict procedures to ensure that:

- Each employee is aware of our policy on food safety and receives appropriate training, instruction and supervision commensurate with his/her work activity, to enable him/her to operate to defined standards.
- Our activities are monitored against the requirements of current legislation and this HACCP Plan.
- Customer satisfaction is guaranteed at all times.

A.4.2 HACCP Team members

	Position	Skills
Team leader	QA/Technical Manager	QA and Operational Hygiene
Team members	Factory Manager	Personnel and Operational Planning
	Kill Floor Manager	Management and Production
	Boning Hall Manager	Management and Production
	Chief Engineer	Maintenance and Planning

A.4.3 Terms of reference: cattle slaughter and dressing

Beef processing from the delivery of animals to the despatch of carcasses (beef quarters) to the butchery operation, with the objective of minimizing:

This case study is an example to illustrate the HACCP approach and is provided without any liability whatsoever in its application and use.

- the number of bacteria on carcasses;
- the risk of contamination with Specified Risk Materials.*

This case study includes the removal and processing/disposal of offal from the beef carcass. Carcasses are despatched without packaging, distributed and stored under chilled conditions until butchery and further processing. This is also carried out on site, in addition to external customers, and is covered by a separate HACCP Plan.

*Specified Risk Materials (SRM) under UK legislation include the head, lungs, spinal cord, spinal cord membrane, intestine, spleen and stomach lining. These materials are required to be removed from the human food chain and destroyed due to the potential presence of BSE infective material which may be a risk to human health.

A.4.4 Process Flow Diagram

Figures A.4.1a–c outline the process, while Figure A.4.1d shows the SRM disposal process. CCP locations have been marked on the relevant diagrams.

A.4.5 Hazard analysis and CCP identification

A full hazard analysis was carried out by the team and CCPs were identified using the Codex decision tree (Codex, 1997). The HACCP Control Chart was used for both hazard analysis and CCP day-to-day management requirements established by the HACCP Team (Table A.4.1). As can be seen from Table A.5, the monitoring, corrective action and responsibility columns are only completed for CCPs.

A.4.6 HACCP Plan implementation

Before implementation, the process operators were all given training covering HACCP and the hygiene requirements relevant to their operation. As CCP monitoring and corrective action were designed to be carried out largely by the QC Team and Management, these individuals were given further specific training in the relevant area. A number of specific HACCP monitoring sheets were developed in addition to the existing process monitoring sheets, which were extended to include some CCPs (see examples in Tables A.4.2 and A.4.3 below). Against this background the HACCP Plan has been successfully implemented and will be reviewed and verified according to the site verification plan covering all HACCP activities.

This case study is an example to illustrate the HACCP approach and is provided without any liability whatsoever in its application and use.

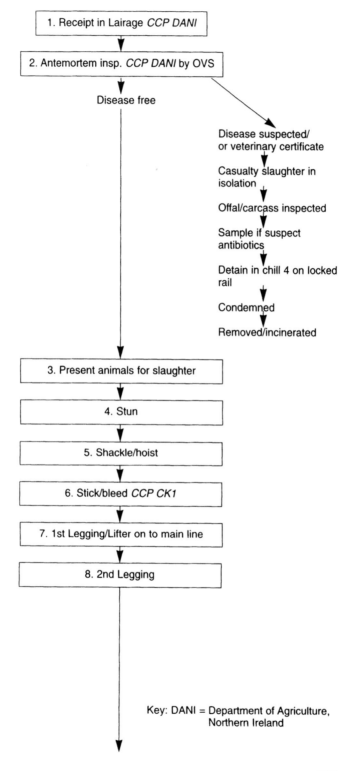

1. Receipt in Lairage *CCP DANI*

2. Antemortem insp. *CCP DANI* by OVS

Disease free

Disease suspected/
or veterinary certificate

Casualty slaughter in
isolation

Offal/carcass inspected

Sample if suspect
antibiotics

Detain in chill 4 on locked
rail

Condemned

Removed/incinerated

3. Present animals for slaughter

4. Stun

5. Shackle/hoist

6. Stick/bleed *CCP CK1*

7. 1st Legging/Lifter on to main line

8. 2nd Legging

Key: DANI = Department of Agriculture,
Northern Ireland

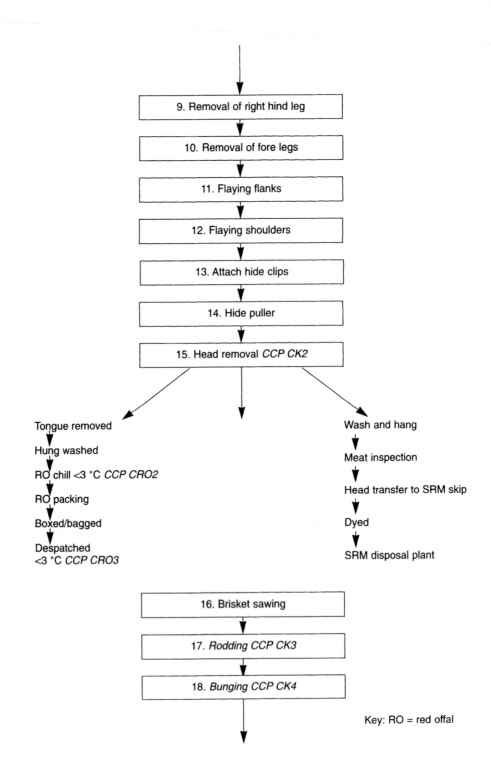

9. Removal of right hind leg

10. Removal of fore legs

11. Flaying flanks

12. Flaying shoulders

13. Attach hide clips

14. Hide puller

15. Head removal *CCP CK2*

Tongue removed
Hung washed
RO chill <3 °C *CCP CRO2*
RO packing
Boxed/bagged
Despatched
<3 °C *CCP CRO3*

Wash and hang
Meat inspection
Head transfer to SRM skip
Dyed
SRM disposal plant

16. Brisket sawing

17. *Rodding CCP CK3*

18. *Bunging CCP CK4*

Key: RO = red offal

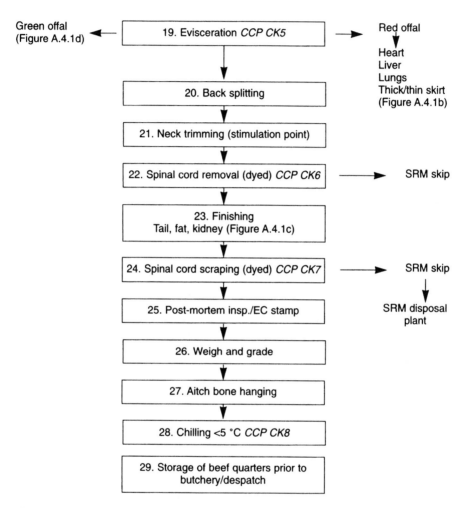

Figure A.4.1 HACCP Flow Diagram (a) of cattle slaughter (for SRM/microbiological hazards);

This case study is an example to illustrate the HACCP approach and is provided without any liability whatsoever in its application and use.

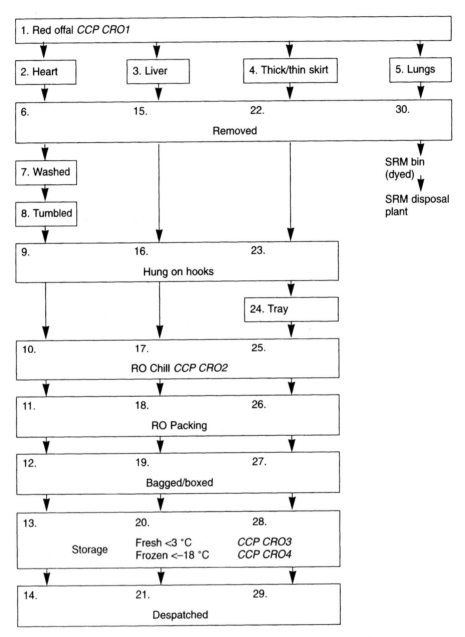

Figure A.4.1 (b) of cattle red offal (for SRM/microbiological hazards);

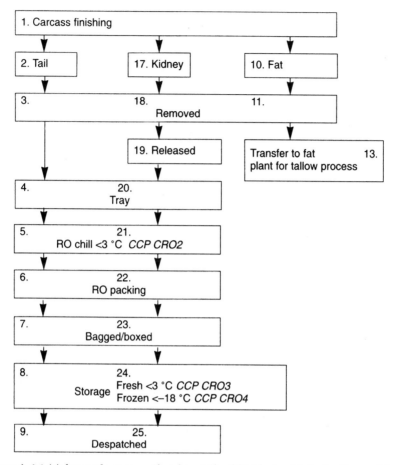

Figure A.4.1 (c) for cattle carcass 'finishing' (for SRM/microbiological hazards).

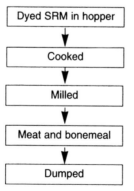

Figure A.4.1 (d) SRM processing – this diagram covers the disposal process for SRM material from the cattle carcass. This is performed in a completely separate processing plant, which has dyed SRM as its raw material. The head, spinal cord, spinal cord scrapings, lungs and all green offal (except the rumen, which is dumped) are processed in this way.

Table A.4.1 HACCP Control Chart. (a) Cattle slaughter for SRM and microbial hazards

Process step	Hazard	Preventative measure	Critical Limits	CCP No.	Monitoring procedure/frequency	Corrective action	Responsibility
001 Reception into lairage	Accepting animals unfit for slaughter	Ear tag number checked on DANI health computer. Auditing of suppliers for compliance with FQAS requirements by Foyle Meats Ltd	100% Compliance	DANI	DANI check all ear tag numbers on health computer/100% inspection	Animal detained until identity is confirmed	DANI
002 Antemortem Insp by OVS	Accepting animals unfit for slaughter	Antemortem inspection	100% compliance	DANI	Antemortem inspection carried out by OVS/100% inspection	Animal is rejected for slaughter if judged to be unfit	DANI
003 Presenting animal for slaughter	Contamination with faecal matter in lairage	Cleaning at the appropriate time	100% compliance	–			
004 Stun	Contamination with faecal matter in cattle run	Cleaning at the appropriate time	100% compliance	–			
005 Shackle/hoist	N/A	N/A		–			

Process step	Hazard	Preventative measure	Critical Limits	CCP No.	Monitoring procedure/frequency	Corrective action	Responsibility
006 Stick/bleed	Microbial contamination as a result of contact with the hide	Training of staff in the correct procedure	100% Compliance	CK1	Kill Floor Manager to carry out visual checks for conformity/hourly audit	Oversight to be brought to the attention of operatives	Kill Floor Manager
	Microbiol contamination as a result of dirty knives	Training of staff in dual knife technique	100% Compliance		QC Dept carry out hourly Audit	Oversight to be brought to the Kill Floor Manager	QC Dept
007 1st Legging	Microbial contamination as a result of contact with dirty knives	Training of staff in the correct procedure for knive sterilization	100% Compliance	—			
	Microbial contamination as a result of contact with dirty aprons/ gloves/ gauntlets	Training of staff in the importance of regular aprons/gloves/gauntlets washing	100%	—			

Table A.4.1 continued

Process step	Hazard	Preventative measure	Critical Limits	CCP No.	Monitoring procedure/frequency	Corrective action	Responsibility
008 2nd Legging	Microbial contamination as a result of contact with dirty knives	Training of staff in the correct procedure for knife sterilization	100% Compliance	–			
	Microbiol contamination as a result of contact with dirty aprons/gloves/gauntlets	Training of staff in the importance of regular aprons/gloves/gauntlets washing	100% Compliance				
009 Removal of right hind leg	Microbial contamination as a result of contact with dirty hock cutter	Training of staff in the correct procedure for hock cutter sterilization	100% Compliance	–			
	Microbial contamination as a result of contact with dirty aprons/gloves/gauntlets	Training of staff in the importance of regular aprons/gloves/gauntlets washing	100% Compliance	–			

Process step	Hazard	Preventative measure	Critical Limits	CCP No.	Monitoring procedure/frequency	Corrective action	Responsibility
0010 Removal of fore legs	Microbial contamination as a result of contact with dirty hock cutter	Training of staff in the correct procedure for hock cutter sterilization	100% Compliance	–			
	Microbiol contamination as a result of contact with dirty aprons/ gloves/ gauntlets	Training of staff in the importance of regular aprons/gloves/gauntlets washing	100% Compliance	–			
0011 Flaying flanks	Microbial contamination as a result of contact with dirty knives	Training of staff in the correct procedure for knife sterilization	100% Compliance	–			
	Microbial contamination as a result of contact with dirty aprons/ gloves/ gauntlets	Training of staff in the importance of regular aprons/gloves/gauntlets washing	100% Compliance	–			

Table A.4.1 *continued*

Process step	Hazard	Preventative measure	Critical Limits	CCP No.	Monitoring procedure/frequency	Corrective action	Responsibility
0012 Flaying shoulder	Microbial contamination as a result of contact with dirty knives	Training of staff in the correct procedure for knife sterilization	100% Compliance	–			
	Microbiol contamination as a result of contact with dirty aprons/gloves/gauntlets	Training of staff in the importance of regular aprons/gloves/gauntlets washing	100% Compliance	–			
0013 Hide clips	Microbial contamination as a result of contact with dirty hide	Training of staff in the correct procedure for the use of hide clips	100% Compliance	–			
	Microbial contamination as a result of contact with dirty aprons/gloves/gauntlets	Training of staff in the importance of regular aprons/gloves/gauntlets washing	100% Compliance	–			

Process step	Hazard	Preventative measure	Critical Limits	CCP No.	Monitoring procedure/frequency	Corrective action	Responsibility
0014 Hide puller	Microbial contamination as a result of contact with dirty hide	Training of staff in the correct procedure for operating hide puller	100% Compliance	–			
	Microbiol contamination as a result of contact with dirty aprons/gloves/gauntlets	Training of staff in the importance of regular aprons/gloves/gauntlets washing	100% Compliance	–			
0015 Head removal	Cross-contamination of SRM and non-SRM	Staff training in the procedure for the effective removal, separation and dying of heads	100% Compliance	CK2	Kill Floor Manager to carry out visual checks conformity/hourly	Oversight to be brought to the attention of operatives. Quarantine and reinspect	Kill Floor Manager
	Microbial contamination as a result of contact with dirty knives	Training of staff in the correct procedure for knife sterilization	100% Compliance	–	QC Dept carry out hourly audit	Oversight to be brought to the attention of the Kill Floor Manager. Quarantine and reinspect	QC Dept

Table A.4.1 *continued*

Process step	Hazard	Preventative measure	Critical Limits	CCP No.	Monitoring procedure/frequency	Corrective action	Responsibility
Head removal cont.	Microbial contamination as a result of contact with dirty aprons/gloves/gauntlets	Training of staff in the importance of regular aprons/gloves/gauntlets washing	100% Compliance	–			
0016 Brisket saw	Microbial contamination as a result of contact with dirty brisket saw	Training of staff in the correct procedure for operating brisket saw and sterilization	100% Compliance	–			
	Microbial contamination as a result of contact with dirty aprons/gloves/gauntlets	Training of staff in the importance of regular apron/gloves/gauntlets washing	100% Compliance	–			

Process step	Hazard	Preventative measure	Critical Limits	CCP No.	Monitoring procedure/frequency	Corrective action	Responsibility
0017 Rodding	Microbial contamination as a result of spillage of the rumen stomach content	Staff training in the correct procedure for sealing the oesophagus	100% Compliance	CK3	Kill Floor Manager to carry out visual checks for conformity/hourly	Oversight to be brought to the attention of operatives	Kill Floor Manager
	Microbial contamination as a result of contact with rodder	Staff training in the correct procedure for sterilization of the rodder between carcasses	100% Compliance	–	QC Dept carry out daily audit	Oversight to be brought to the attention of the Kill Floor Manager	QC Dept
0018 Bunging	Microbial contamination of the carcass and operatives as a result of faeces from rectum	Staff training in the correct procedure for tying and bagging the anus	100% Compliance	CK4	Kill Floor Manager to carry out visual checks for conformity/hourly	Oversight to be brought to the attention of operatives	Kill Floor Manager
					QC Dept carry-out daily audit	Oversight to be brought to the attention of the Kill Floor Manager	QC Dept

Table A.4.1 *continued*

Process step	Hazard	Preventative measure	Critical Limits	CCP No.	Monitoring procedure/frequency	Corrective action	Responsibility
0019 Evisceration	Microbial contamination of the carcass and operatives as a result of spillage from the large intestine	Staff training in the correct procedure for knife sterilization	100% Compliance	CK5	Kill Floor Manager to carry out visual checks for conformity/hourly	Oversight to be brought to the attention of operatives. Quarantine and reinspect	Kill Floor Manager
		Staff training in the correct procedure for removal of the large intestine	100% Compliance	–	QC Dept carry out daily audit	Oversight to be brought to the attention of the Kill Floor Manager. Quarantine and reinspect	QC Dept
0020 Back splitting	Microbial contamination as a result of contact with dirty saw	Training of staff in the correct procedure for operating saw and sterilization	100% Compliance	–			
	Microbial contamination as a result of contact with dirty aprons/gloves/gauntlets	Training of staff in the importance of regular aprons/gloves/gauntlets washing	100% Compliance	–			

Process step	Hazard	Preventative measure	Critical Limits	CCP No.	Monitoring procedure/frequency	Corrective action	Responsibility
Back splitting cont.	Incorrect sawing of back not allowing removal of the spinal cord	Staff training in the procedure for back splitting	100% Compliance	–			
0021 Neck trimming	Microbial contamination as a result of contact with dirty knives	Training of staff in the correct procedure for knife sterilization	100% Compliance	–			
	Microbial contamination as a result of contact with dirty aprons/ gloves/ gauntlets	Training of staff in the importance of regular aprons/gloves/gauntlets washing	100% Compliance	–			

Table A.4.1 *continued*

Process step	Hazard	Preventative measure	Critical Limits	CCP No.	Monitoring procedure/frequency	Corrective action	Responsibility
0022 Spinal cord removal (dyed)	Microbial contamination as a result of contact with dirty knives	Training of staff in the correct procedure for knife sterilization	100% Compliance	CK6	Kill Floor Manager to carry out visual checks for conformity/hourly	Oversight to be brought to the attention of the operatives. Quarantine and reinspect. Carcasses removed to detained chill until spinal cord is correctly removed	Kill Floor Manager
	Microbial contamination as a result of contact with dirty aprons/gloves/gauntlets	Training of staff in the importance of regular aprons/gloves/gauntlets washing	100% Compliance				
	Spinal cord has not been removed correctly	Staff training in the correct procedure for removal and dying of the spinal cord	Complete removal of the spinal cord	CK6	QC Dept carry out daily audit Procedure FQC 014	Oversight to be brought to the attention of the Kill Floor Manager. Quarantine and reinspect	QC Dept
0023 Finishing tail, fat, kidney	Microbial contamination as a result of contact with dirty knives	Training of staff in the correct procedure for knife sterilization	100% Compliance	–			

Process step	Hazard	Preventative measure	Critical Limits	CCP No.	Monitoring procedure/frequency	Corrective action	Responsibility
Finishing tail, fat, kidney cont.	Microbial contamination as a result of contact with dirty aprons/gloves/gauntlets	Training of staff in the importance of regular aprons/gloves/gauntlets washing	100% Compliance	–			
0024 Spinal cord scraping (dyed)	Microbial contamination as a result of contact with dirty knives	Training of staff in the correct procedure for knife sterilization	100% Compliance	CK7	Kill Floor Manager to carry out visual checks for conformity/hourly	Oversight to be brought to the attention of operatives. Quarantine and reinspect Carcasses removed to detained chill until spinal cord membrane is correctly removed	Kill Floor Manager
	Microbial contamination as a result of contact with dirty aprons/gloves/gauntlets	Training of staff in the importance of regular aprons/gloves	100% Compliance	–			
	Spinal cord membrane has not been removed correctly	Staff training in the correct procedure for the removal and dying of the spinal cord membrane	100% Compliance	CK7	QC Dept carry out daily audit Procedure FQC 014	Oversight to be brought to the attention of the Kill Floor Manager. Quarantine and reinspect	QC Dept

Table A.4.1 *continued*

Process step	Hazard	Preventative measure	Critical Limits	CCP No.	Monitoring procedure/frequency	Corrective action	Responsibility
0025 Post-mortem inspection/EC stamping	Accepting a diseased carcass which is unfit for human consumption	Training of Meat Inspectors in the identification of unfit or diseased carcasses by the DANI	100% Compliance	DANI	Visual inspection of carcasses for the detection of disease and elimination of abnormal carcasses by the DANI Meat Inspectors/100% sampling	Carcasses are detained for additional inspection by the Senior Meat Inspector and rejected where necessary	DANI
0026 Weigh and grade	N/A	N/A		—			
0027 Carcass wash	Non-removal of carcass contamination	Training of staff in the correct procedure for carcass washing to remove contamination	100% Compliance	—			
	Microbial contamination as a result of contact with dirty aprons/gloves/gauntlets	Training of staff in the importance of regular aprons/gloves/gauntlets washing	100% Compliance				

Process step	Hazard	Preventative measure	Critical Limits	CCP No.	Monitoring procedure/frequency	Corrective action	Responsibility
0028 Aitch bone hanging	Microbial contamination as a result of contact with dirty hooks	Training of staff in the correct procedure for hook sanitation	100% Compliance	–			
	Microbial contamination as a result of contact with dirty aprons/gloves/gauntlets	Training of staff in the importance of regular aprons/gloves/gauntlets washing	100% Compliance	–			
0029 Chilling <5 °C and storage prior to despatch or boning	Growth of bacterial flora as a result of temperature being greater than 5 °C	Monitoring temperature using the INENCO system and maintenance of carcass temperature below 5 °C	Temperature must be held below 5 °C	CK8	QC Dept carryout a daily audit and constantly monitor temperature through the INENCO system	Oversight to be brought to the attention of QC Dept who contacts Chief Engineer to correct problem and carcass detained	QC Dept Chief Engineer
	Growth of bacterial flora as a result of over-crowding within the chill	Staff training in the correct procedure for spacing carcasses in the chill area	Carcasses spaced correctly	–	Kill Floor Manager to carry out visual checks for conformity/hourly QC Dept carry out daily audit	Oversight to be brought to the attention of the operatives and wrongly spaced carcasses respaced correctly	Kill Floor Manager QC Dept

Table A.4.1 continued

Process step	Hazard	Preventative measure	Critical Limits	CCP No.	Monitoring procedure/frequency	Corrective action	Responsibility
Chilling <5 °C and storage prior to despatch or boning cont.	Microbial contamination as a result of carcasses coming into contact with chill walls and doors	Staff training in the correct procedure in moving carcasses and effective cleaning and sanitizing of the chill area	100% Compliance	–	Kill Floor Manager to carry out visual checks for conformity/hourly	Oversight to be brought to the attention of operatives Quarantine and reinspect	Kill Floor Manager
					QC Dept carry out daily audit Procedure FQC 004	Oversight to be brought to the attention of the Kill Floor Manager. Quarantine and reinspect	QC Dept
0030 Cleaning	Growth of bacterial flora as a result of microbial contamination and unhygienic conditions within the factory surfaces	Training of staff in the correct procedures for cleaning as laid out in the cleaning manual for each factory area	Carry out cleaning effectively and efficieny	CK9	QC Dept will undertake the training of cleaning team in cleaning methods and techniques	Oversight to be brought to the attention of the QC Manager and cleaners retrained	QC Dept
		Cleaning of the factory on an ongoing basis and after production cleaning shift	100% Compliance	–	QC Dept will carry out cleaning audit/daily basis Procedure FQC 003	Oversight to be brought to the attention of the QC Manager Areas which are not satisfactory will be recleaned	QC Dept Charge Hand

Process step	Hazard	Preventative measure	Critical Limits	CCP No.	Monitoring procedure/frequency	Corrective action	Responsibility
Cleaning cont.		Microbiological analysis of the hygiene standards on factory surfaces	TVC < 100 cfu/cm^2 No coliforms present	–	QC Dept will carry out microbial analysis of hygiene of factory areas by aseptic swabbing/ weekly basis Procedure FQC 011 (1)	Oversight to be brought to the attention of the QC Manager In areas which are not satisfactory the cleaning schedule will be reviewed	QC Dept

(b) Cattle red offal for SRM and microbial hazards

Process step	Hazard	Preventative measure	Critical Limits	CCP No.	Monitoring procedure/frequency	Corrective action	Responsibility
001 Red offal removed from the carcass	Contamination as a result of contact with dirty knife	Training of staff in the correct procedure for knife sterilization	100% Compliance	—			
	Contamination as a result of contact with dirty aprons/gloves/gauntlets washing	Training of staff in the importance of regular aprons/gloves/gauntlets washing					
002 Red offal chilling <3 °C	Growth of bacterial flora as a result of temperature being greater than 3 °C	Monitoring of temperature using the INENCO system and maintenance of temperature below 3 °C	Temperature must be held below 3 °C	CR01	QC Dept carry out a daily audit and constantly monitor temperature through the INENCO system	If an oversight occurs it is brought to the attention of QC Dept Manager who contacts Chief Engineer to correct problem and red offal detained	QC Dept Manager Chief Engineer
003 Red offal packaged	Contamination as a result of contact with dirty aprons/gloves/gauntlets	Training of staff in the importance of regular aprons/gloves/gauntlets washing	100% Compliance	—			

Process step	Hazard	Preventative measure	Critical Limits	CCP No.	Monitoring procedure/frequency	Corrective action	Responsibility
004 Red offal packs chilled	Growth of bacterial flora as a result of temperature being greater than 3 °C	Monitoring of pack temperature and maintenance of temperature below 3 °C	Temperature must be below 3 °C	CR02	Quality Control Dept check red offal packaged temperature Procedure FQC 023/check prior to depatch	If oversight occurs it is brought to the attention of QC Dept Manager who contacts Chief Engineer to correct problem and red offal detained	QC Dept Manager Chief Engineer
005 Red offal packs freeze	Growth of bacterial flora as a result of temperature being greater than −18 °C	Monitoring of pack temperature and maintenance of temperature below −18 °C	Temperature must be below −18 °C	CR03	Quality Control Dept check red offal packaged temperature Procedure FQC 023/check prior to despatch	If oversight occurs it is brought to the attention of QC Dept Manager who contacts Chief Engineer to correct problem and red offal detained	QC Dept Manager Chief Engineer
Cleaning	Growth of bacterial flora as a result of contamination and unhygienic conditions within the factory surfaces	Training of staffing in the correct procedures for cleaning as laid out in the cleaning manual for each factory area Cleaning of the factory an ongoing basis and after production cleaning	Carry out cleaning effectively and efficieny 100% Compliance	CR05	QC Dept will under take the training of cleaning team in cleaning methods and techniques QC Dept will carry out cleaning audit daily Procedure FQC 005	Oversight to be brought to the attention of the QC Manager and cleaners rerained Oversight to be brought to the attention of the QC Manager Areas which are not be recleaned	QC Dept QC Dept Charge Hand

Table A.4.1 *continued*

Process step	Hazard	Preventative measure	Critical Limits	CCP No.	Monitoring procedure/frequency	Corrective action	Responsibility
Cleaning cont.		Microbial analysis of the hygiene standards on factory surfaces	TVC < 100 cfu/cm^2 No coliforms Present		QC Dept will carryout microbial analysis of hygiene of factory areas by aseptic swabbing/ weekly basis/Procedure FQC	Oversight to be brought to the attention of the QC Manager In areas which are not satisfactory the cleaning schedule will be reviewed	QC Dept

Table A.4.2 HACCP monitor sheet – cattle slaughter for SRM and microbial hazards CCPs

Date:
Hourly audit by: Verified by:

CCP/No.   Controlled?   Time:
Job   Y/N

	Hazards/Control	8.00 am	9.00 am	10.00 am	11.00 am	12.00 pm	1.00 pm	2.00 pm	3.00 pm	4.00 pm	Action by
CK1 Sticking	Dirt from hide into neck when cutting/dual knife system										
CK2 Head removal	Incorrect removal of head/staff training in the correct procedure										
CK3 Rodding	Contamination from gut contents/ sealing oesophagus										
CK4 Bunging	Contamination from gut contents/ tying and bagging										
CK5 Evisceration	Contamination from leaked stomach contents/staff training in the correct procedure										
CK6 Removal spinal cord	Incorrect removal of spinal cord/ staff training in the correct procedure										
CK7 Removal spinal cord membrane	Incorrect removal of spinal cord membrane/staff training in the correct procedure										
CK8 Chill	Microbial growth/ chill 5 °C										
CK9 Cleaning	Microbial growth/ cleaning										

Good Manufacturing Practice (GMP)

Job	Must do	Y/N
Flaying flanks/shoulders	Clip hide back well, the outside hide must not touch the clean meat	
Carcass washer	Wash off dirt, blood, rumen	
Trimming	Remove thymus from oesophagus, remove contamination	

Table A.4.3 Monitoring of carcass pH and spinal cord removal (example of a monitoring sheet including CCP)

1. pH Measurements

Procedure: All cattle in chill at start of shift checked (see FQC 014 for method)

Date: Signature:

Customer	DOK	Highest value	Lowest value

pHs' Outside customer specification

Customer	DOK	Value	Amount outside spec.	Action Kill No.

2. Spinal cord checks

Procedure: At least 10% of carcasses checked at same time as pH measurement (see FQC 014 for method)

Spinal cord effectively removed   Yes/No

Approx. no. checked	DOK	Time of kill	Action

A.5 Case study – Meals on Wheels, a safe food system (community catering)

Denise Worsfold, Cardiff Business School, University of Wales, 66 Park Place, Cardiff CF1 3AT, Wales

A.5.1 Introduction

The Women's Royal Voluntary Service (WRVS) works independently and in co-operation with the Health and Local Authorities to provide elderly and disabled people with 'community meals'. The service operates 2200 Meals on Wheels (MoW) schemes and delivers 1.5 million meals annually. It is mainly staffed by volunteers, who work in local food projects, each run by a project organizer. District and county food managers supervise the work of local projects and report to divisional food managers and a director of the food service. Hot meals for daily delivery are produced from fresh ingredients cooked on the premises, or are regenerated from frozen food.

A.5.2 HACCP Team members

- Director of Food Service
- Divisional Food Manager
- District Food Organizer
- MoW Project Leader.

All the above have wide experience of the Meals on Wheels (MoW) Service at different levels and have been involved with food safety training. The team was completed by the Home Authority representative with experience of food safety legislation and a Food Safety Advisor with research experience of HACCP in small-scale operations.

A.5.3 Terms of reference

Start and finish point of the study: purchase of frozen individual meal to service.
    Hazards to be considered: microbial, physical.
    Consumers: frail, elderly.

This case study is an example to illustrate the HACCP approach and is provided without any liability whatsoever in its application and use.

A.5.4 Product description

A manufactured individual meal consisting of meat/poultry/fish together with potato/rice/pasta and vegetables. The product is regenerated from frozen and transported hot to the consumer in his or her own home.

A.5.5 Process Flow Diagram

Step 1	Purchase of frozen meals
Step 2	Receipt
Step 3	Storage
Step 4	Regeneration
Step 5	Loading transport containers
Step 6	Transport
Step 7	Service

A.5.6 Hazard analysis

(a) Step 1 Purchase of frozen meals

Hazards and causes:

• Unreliable and unhygienic suppliers are used with the likelihood that the food becomes contaminated.

Control measures:

• Use reliable and hygienic suppliers.

(b) Step 2 Receipt

Hazards and causes:

• Product packaging is damaged, so the food becomes contaminated with pathogens or foreign bodies, such as string, staples and wood splinters.
• The food is out of date, possible increase in contamination.
• Delivery temperature is too high, allowing any pathogens to increase in number.

Control measures:

• Reject deliveries with damaged packaging.
• Check the date stamps.
• Check the temperature of delivered frozen products. Must be no higher than –15 °C.

This case study is an example to illustrate the HACCP approach and is provided without any liability whatsoever in its application and use.

(c) Step 3 Storage

Hazards and causes:

- Frozen deliveries are left at room temperature for too long. This could lead to an increase in contamination.
- Frozen products are stored in freezers which are too warm. This could allow some pathogens to multiply.
- No stock rotation is observed, products get out of date, leading to an increase in contamination.
- The product packaging is inadequate/damaged leading to the food becoming contaminated.

Control measures:

- Store frozen products in freezers immediately.
- Check the temperature of the freezer, frozen products food must be stored at –18 °C.
- Stack stored foods to allow the first in to be used before the last in. Avoid overstocking.
- Check that the packaging of food is clean and adequate.

(d) Step 4 Regeneration

Hazards and causes:

- Product is not regenerated to a safe temperature and bacteria may survive.

Control measures:

- Check that the regenerated meals reach an internal temperature of at least 85 °C.

(e) Step 5 Loading transport containers

Hazards and causes:

- Delays allowing the temperature of the regenerated products to fall.

Control measures:

- Ensure that the transport containers are preheated.
- Ensure that the regenerated meals are loaded quickly into the containers and the lids are replaced promptly.
- Ensure that the delivery personnel collect the loaded transport containers promptly.

This case study is an example to illustrate the HACCP approach and is provided without any liability whatsoever in its application and use.

(f) Step 6 Transport

Hazards and causes:

- The temperature of the hot food is too low and could lead to an increase in contamination.
- Transport containers are dirty, allowing the food to become contaminated.
- Regenerated food is not adequately protected during transport, allowing cross-contamination to occur.

Control measures:

- Limit the length of the delivery round to less than 90 minutes.
- Ensure that delivery staff remove the meal at the last possible moment and that the lid is replaced.
- Transport containers must be clean.
- Regenerated meals are packed carefully to avoid spillage and package damage.

(g) Step 7 Service

Hazards and causes:

- If service is delayed, there may be an increase in contamination.
- If hands of delivery staff are not clean, cross-contamination is possible.

Control measures:

- Delivery staff to wrap meal during transit from vehicle to house.
- Meal may not be left on doorstep or with a neighbour.
- Temperature check of sample meal to be taken at end of delivery round on regular basis.
- Delivery staff to maintain high standard of personal hygiene.

All control measures were already in operation and identified in a code of practice. Regeneration and delivery personnel have all undertaken in-house food safety training. New, more comprehensive monitoring record sheets had to be designed and tested.

A.5.7 Identification of Critical Control Points

The Critical Control Points were identified as regeneration and the loading and transport steps that followed. The team considered whether a loss of control at each step would lead to a 'significant' health risk. Where this was considered to be the case,

---

This case study is an example to illustrate the HACCP approach and is provided without any liability whatsoever in its application and use.

Critical Control Points were identified. CCP Decision Trees were not used.

A.5.8 Description of how CCPs are managed

Food temperature is monitored at regeneration and at the point of delivery. In addition, food temperature and condition are monitored at delivery and appliance temperature monitored during storage. Details of all the monitoring procedures are given on the HACCP Plan sheet (Table A.5.1).

A.5.9 Implementation and maintenance

The process Flow Chart was verified by members of the HACCP Team in a sample of projects. New monitoring forms were tested in a pilot study, improved and training given in their completion. A simple system of documentation, the Safe Food System, consisting of a user's manual and two checklists was developed, tested in a pilot study and the format improved. MoW projects are required to complete the checklists, thereby confirming that they have identified the process hazards and affirming that they are using appropriate controls and monitoring methods.

Food managers attended a series of half-day workshops, where they were introduced to the principles of hazard analysis and to the Safe Food System. They were advised on suitable monitoring methods and on training for staff to carry out the necessary checks and take appropriate corrective actions.

Food managers have introduced the Safe Food System to the project leaders of the highest-risk projects in their areas. They are now in the process of assisting them in completing their hazard analyses. The documentation is customized for each unit and then one copy is returned to the food manager for checking and filing. Food and project managers regularly scrutinize monitoring sheets to ensure that the system is running as planned.

For the Meals on Wheels operation the documentation includes a standards sheet (Figure A.5.1) which is inserted into the front of the kitchen diary, and a self-adhesive monitoring form (Figure A.5.2) which is attached to the relevant diary pages for each monitoring (operational) day.

This case study is an example to illustrate the HACCP approach and is provided without any liability whatsoever in its application and use.

Table A.5.1 HACCP Plan: Meals on Wheels

Product/Process: *Meals on wheels*

Regenerated from frozen, delivered hot

No.	Process stage	Hazard	Method of control or prevention	CCP?	Specification/ Critical Limits	Monitoring Procedure	Monitoring Frequency	Corrective actions	Responsibility
1	Purchase of frozen meals	Unreliable, unhygienic suppliers, food becomes contaminated	Use reliable and hygienic suppliers						Contract Manager
2	Receipt	Product packaging damaged, food contaminated	Reject deliveries with damaged packaging			Visual checks	Each delivery	Reject delivery if out of specification	Project Leader/Cook
		Product is out of date, increase in contamination	Check date stamps			Visual checks	Each delivery		Project Leader/Cook
		Delivery temperature is too high, increase in contamination	Check temperature of frozen products, must be no higher than −15 °C		Below −15 °C	Temp. check Record	Each delivery	Reject delivery if out of specification	Project Leader/ Cook
3	Storage	Growth of pathogens, toxins on high-risk food Further contamination	Store products in freezer promptly Store at correct temperature		Below −18 °C	Temp. check Record	Daily	Adjust thermostat, recheck Report for repair	Cook

Product/Process: _Meals on wheels_

Process stage	Hazard	Method of control or prevention	CCP?	Specification/ Critical Limits	Monitoring		Corrective actions	Responsibility
					Procedure	Frequency		
No. 3 Storage cont.		Use by recommended date Check packaging			Visual checks			Cook
4 Regeneration	Survival of pathogens	Heat until temperature of meal is 85 °C or above	✓	85 °C or above	Temperature check Record	Each production run	Do not remove meals from oven until products reach 85 °C	Cook
5 Loading of transport containers	Temperature of regenerated food falls	Load promptly replace lid	✓	Preheated load within 10 min	Supervisory checks	Each production run	Check food temp. If there is a delay, reheat if necessary	Cook
6 Transport	Temperature of food is too low, could be increase in contamination	Keep lid on box Remove meal from box at last minute	✓	Meal on delivery to client must be above 63 °C	Temp. check of sample meal	Weekly/ spot checks	Return meals below 63 °C to kitchen. Report Provide alternative food to client	Delivery staff

Table A.5.1 *continued*

Product/Process: _Meals on wheels_

No.	Process stage	Hazard	Method of control or prevention	CCP?	Specification/ Critical Limits	Monitoring		Corrective actions	Responsibility
						Procedure	Frequency		
	Transport cont.	Further contamination	Limit round length to 90 min or less			Record time of delivery to last client	Each delivery round	If delivery time too long, report to round organiser	
			Ensure delivery boxes are clean		Boxes to be clean and well maintained	Visual checks	Each delivery round		
7	Service	If delayed, increase in contamination	Lone drivers should summon help if delayed by vehicle breakdown, client emergency Round should be continued if possible		Use insulated bag for multi-site delivery Wrap individual meals Do not leave meal on doorstep or with neighbour	Problems to be reported to Round Organizer	When necessary	Arrange alternative delivery, e.g. taxi	Delivery staff Round organizer

Product/Process: *Meals on wheels*

Process stage		Hazard	Method of control or prevention	CCP.?	Specification/ Critical Limits	Monitoring		Corrective actions	Responsibility
						Procedure	Frequency		
No.	Service cont.	Further contamination	Good personal hygiene		All delivery staff trained in food hygiene	Supervisory checks	Ongoing	Retrain staff	Project or Round Organizer

DELIVERIES – CHECKS AND TARGETS			
	CONDITION	**SHELF-LIFE**	**TEST**
WHEN	On receipt	On receipt	On receipt
HOW	Visual	Visual	Temperature probe
TARGET	Good, pack intact	Correct date stamp	Chilled food must be less than **8 ˚C** Frozen food must be no higher than **–15 ˚C** Hot, cooked food must be no less than **68 ˚C**
CORRECTIVE ACTION	Reject bad deliveries	Reject bad deliveries	Reject bad deliveries

FOOD STORAGE – CHECKS AND TARGETS			
TEST	**TEMP. OF ALLIANCE**	**CROSS-CONTAMINATION**	**SHELF-LIFE**
WHEN	Fridge – daily Freezer – daily	Daily	Daily
HOW	Temp. probe	Visual	Visual
CORRECTIVE ACTION	Adjust or repair	Separate food, clean appliance	Discard

CLEANING – CHECKS AND TARGETS			
TEST	**EQUIPMENT**	**SURFACES**	**CLOTHES**
WHEN	Between and at end of production	Between and at end of production	Start and at end of production
HOW	Visual	Visual	Visual
TARGET	Clean (disinfect where necessary)	Clean (disinfect where necessary)	Clean (disinfect regularly)
CORRECTIVE ACTION	Reclean, check cleaning schedule	Reclean	Discard, replace

COOKING AND REGENERATION – CHECKS AND TARGETS	
TEST	**TEMPERATURE OF COOKED 'HIGH-RISK' FOOD**
WHEN	At the end of cooking/regeneration
HOW	Record time and oven temperature. Check with temp. probe
TARGET	Cook to centre temperature above **75 ˚C**
CORRECTIVE ACTION	Keep cooking until confident that the temperature is reached

NOTES
1. Record the above information for the main course of meat, offal, fish or eggs. Record the information for milk puddings like rice, semolina and custard.
2. Confirm that the cooking time and oven temperature are satisfactory by checking periodically with a digital thermometer.
3. Ensure that the temperature probe is cleaned and disinfected between uses.
4. Record any incidents where a test has failed and what action was taken.

Figure A.5.1 Safe Food System – monitoring standards sheet.

KEY	
Ck/T	Cooking Time
Ov/S	Oven Setting
Con/Store	Condition of Store
Con/Goods	Condition of Goods
Date st.	Date stamps
MoW	Meals on Wheels
LC	Lunch Club
RCM	Recommended Cleaning Methods

Figure A.5.1 *continued*

DELIVERY CHECK				REGENERATION CHECK				
Supplier	Ok as target	Action	Sig	Main	Ck/T	Ov/S	°C	Sig
FOOD STORAGE CHECK								
Appliance	Actual °C	Action	Sig					
				Sweet				
DRY GOODS								
Con/Store	Con/Goods	Date st.	Sig	NUMBER OF MEALS COOKED				
				Mow		Date		
CLEANING CHECK				LC				
Ok as target using RCM		Sig		Cook/Organizer				
				Helper				

Figure A.5.2 Safe Food System – monitoring record sheet.

This case study is an example to illustrate the HACCP approach and is provided without any liability whatsoever in its application and use.

A.6 Case study – Chocolate Mini Eggs

P. Nugent, Cadbury Chocolate Canada Inc., Toronto, Canada

A.6.1 Introduction

One of the first HACCP studies undertaken by this company was the manufacturing of Mini Eggs. This production line represents a typical process in the manufacturing of confectionery products as one of the raw materials, chocolate, is made in another department and transferred by pumps and lines to its usage point. This gives rise to the potential introduction of a number of foreign materials.

A.6.2 HACCP Team members

The HACCP team consisted of the following:
- Quality Assurance Manager
- Microbiologist
- Production Supervisor
- Manager, Process Integration
- Maintenance Supervisor.

A.6.3 HACCP Study terms of reference

To identify the physical, chemical and microbiological hazards associated with this process, with particular attention to the potential for the introduction of foreign material.

A.6.4 Product Description

Mini Eggs consist of a moulded chocolate centre which is panned with layers of gum arabic, candy and colour to achieve a smooth surface.

A.6.5 Process Flow Diagrams

As this is a complex operation, the HACCP Team constructed an overview flow diagram (Figure A.6.1) and a series of detailed flow diagrams covering different process modules (Figure A.6.2).

This case study is an example to illustrate the HACCP approach and is provided without any liability whatsoever in its application and use.

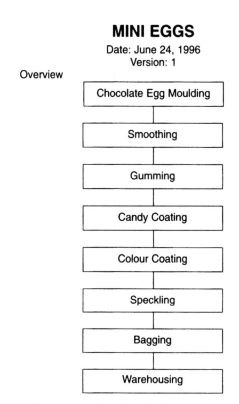

MINI EGGS

Date: June 24, 1996
Version: 1

Overview

Chocolate Egg Moulding

Smoothing

Gumming

Candy Coating

Colour Coating

Speckling

Bagging

Warehousing

Figure A.6.1 Mini Eggs – overview Flow Diagram.

This case study is an example to illustrate the HACCP approach and is provided without any liability whatsoever in its application and use.

MINI EGGS

Chocolate Egg Making (Smart Plant) and Smoothing

Chocolate Supply to Mini Eggs
- Chocolate coating from Choc. dept. pumped to 4th floor storage tank (mild steel with agitator)
- Pumped to pre-tempering tank (mild steel with agitator)

Tempering Unit
- In Aasted tempering unit temper using correct temperature profile (see SOP)

Mini Eggs Moulding
- Smart Moulding Unit, comprising of 2 brass press rollers kept at −13 to −15 °C with Glycol Brine, is used to mould chocolate eggs

---

Mini Eggs Cooling System and smoothing
- Formed mini eggs go through cooling tunnel
- Egg edges shaved off by rolling through revolving drums, shavings then collected and manually dumped into a pre-tempering tank for rework

---

Mini Eggs Conditioning
- Mini eggs are placed in the beige trays on the weighing scale (see SOP)
- Each Mini Eggs tray weighs 20 lb (see SOP)
- Trays are labelled with time/shift/date (production card)
- Stack beige trays with lids on the production floor before panning
- Allow trays to sit at room temp. for 12 hours
- After 12 hours mini eggs are ready for panning

① Chocolate Eggs Conditioned

Figure A.6.2 Mini Eggs – modular Process Flow Diagrams.

This case study is an example to illustrate the HACCP approach and is provided without any liability whatsoever in its application and use.

Preparing Gum Solution

<u>Gum Arabic</u>
- Received in 35 kg Kraft plastic lined bags in 1st floor receiving area
- Stored in the basement cellar (Temp. 9 °C)
- Each bag brought to 4th floor using dolly
- Bags are cut open using knife then emptied into a orange storage bin

---

<u>Weighing Gum Arabic</u>
- Pour hot city tap water (Temp. 60 °C) into a white pail on the weighing scale (see SOP)
- Scoop Gum Arabic into the hot water (see SOP)
- Mix with Hayword Gordon mixer
- Repeat same mixture 3 times in 3 different white pails and place pails on the plastic skids

---

<u>Cooking and mixing Gum Arabic</u>
- Meter in liquid sucrose to cooking kettle from basement bulk tank
- Cook liquid sucrose to 106 °C, check solids reading to 73 (refractometer)
- Allow liquid sugar to stop boiling and cool after cooking
- Add 3 pails of gum mixture to the sugar kettle
- Start kettle mixer and mix for 30 sec maximum (see SOP)

<u>Conditioning Gumming Solution</u>
- After cooling, hold Arabic Gum mixture for 11 hours before using

② Conditioned Gum

Figure A.6.2 *continued*

This case study is an example to illustrate the HACCP approach and is provided without any liability whatsoever in its application and use.

Gumming the Eggs

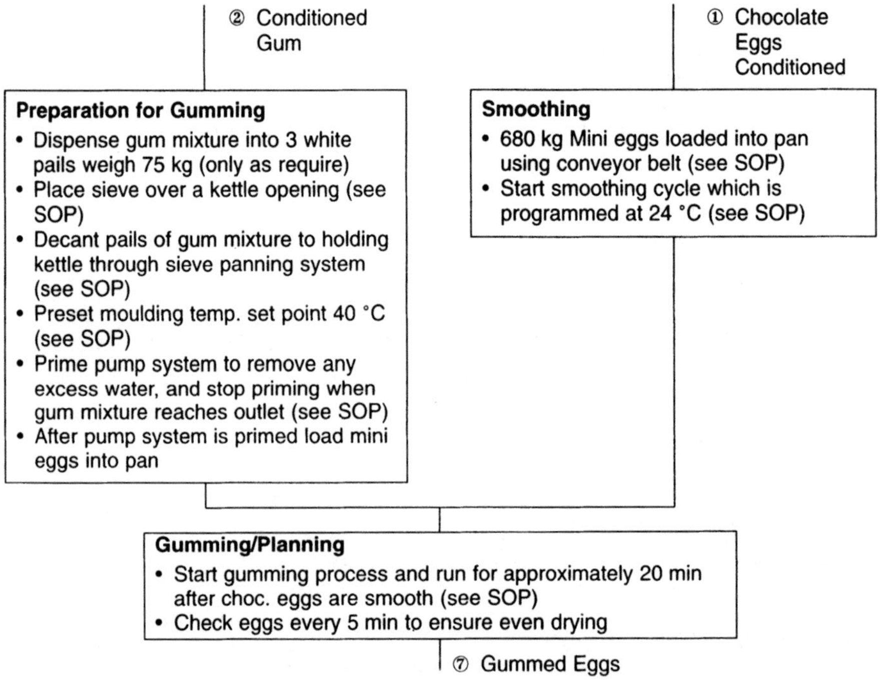

② Conditioned
Gum

① Chocolate
Eggs
Conditioned

Preparation for Gumming
- Dispense gum mixture into 3 white pails weigh 75 kg (only as require)
- Place sieve over a kettle opening (see SOP)
- Decant pails of gum mixture to holding kettle through sieve panning system (see SOP)
- Preset moulding temp. set point 40 °C (see SOP)
- Prime pump system to remove any excess water, and stop priming when gum mixture reaches outlet (see SOP)
- After pump system is primed load mini eggs into pan

Smoothing
- 680 kg Mini eggs loaded into pan using conveyor belt (see SOP)
- Start smoothing cycle which is programmed at 24 °C (see SOP)

Gumming/Planning
- Start gumming process and run for approximately 20 min after choc. eggs are smooth (see SOP)
- Check eggs every 5 min to ensure even drying

⑦ Gummed Eggs

Figure A.6.2 *continued*

This case study is an example to illustrate the HACCP approach and is provided without any liability whatsoever in its application and use.

Candy Coating

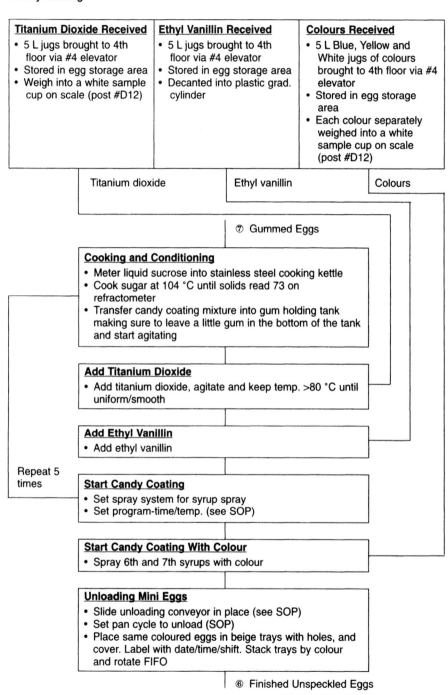

Titanium Dioxide Received	Ethyl Vanillin Received	Colours Received
• 5 L jugs brought to 4th floor via #4 elevator • Stored in egg storage area • Weigh into a white sample cup on scale (post #D12)	• 5 L jugs brought to 4th floor via #4 elevator • Stored in egg storage area • Decanted into plastic grad. cylinder	• 5 L Blue, Yellow and White jugs of colours brought to 4th floor via #4 elevator • Stored in egg storage area • Each colour separately weighed into a white sample cup on scale (post #D12)

Titanium dioxide   Ethyl vanillin   Colours

⑦ Gummed Eggs

Cooking and Conditioning
- Meter liquid sucrose into stainless steel cooking kettle
- Cook sugar at 104 °C until solids read 73 on refractometer
- Transfer candy coating mixture into gum holding tank making sure to leave a little gum in the bottom of the tank and start agitating

Add Titanium Dioxide
- Add titanium dioxide, agitate and keep temp. >80 °C until uniform/smooth

Add Ethyl Vanillin
- Add ethyl vanillin

Repeat 5 times

Start Candy Coating
- Set spray system for syrup spray
- Set program-time/temp. (see SOP)

Start Candy Coating With Colour
- Spray 6th and 7th syrups with colour

Unloading Mini Eggs
- Slide unloading conveyor in place (see SOP)
- Set pan cycle to unload (SOP)
- Place same coloured eggs in beige trays with holes, and cover. Label with date/time/shift. Stack trays by colour and rotate FIFO

⑥ Finished Unspeckled Eggs

Figure A.6.2 *continued*

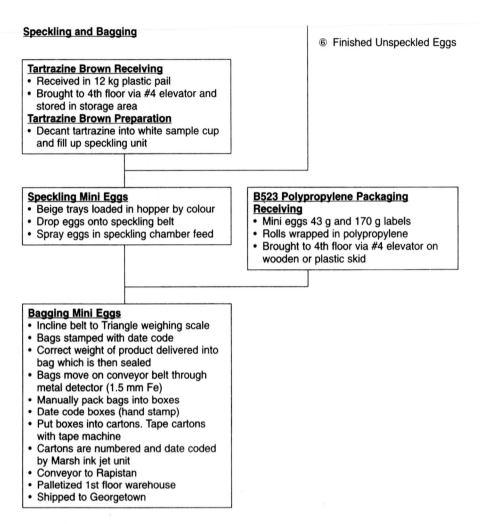

Speckling and Bagging

⑥ Finished Unspeckled Eggs

Tartrazine Brown Receiving
• Received in 12 kg plastic pail
• Brought to 4th floor via #4 elevator and
  stored in storage area
Tartrazine Brown Preparation
• Decant tartrazine into white sample cup
  and fill up speckling unit

Speckling Mini Eggs
• Beige trays loaded in hopper by colour
• Drop eggs onto speckling belt
• Spray eggs in speckling chamber feed

B523 Polypropylene Packaging Receiving
• Mini eggs 43 g and 170 g labels
• Rolls wrapped in polypropylene
• Brought to 4th floor via #4 elevator on
  wooden or plastic skid

Bagging Mini Eggs
• Incline belt to Triangle weighing scale
• Bags stamped with date code
• Correct weight of product delivered into
  bag which is then sealed
• Bags move on conveyor belt through
  metal detector (1.5 mm Fe)
• Manually pack bags into boxes
• Date code boxes (hand stamp)
• Put boxes into cartons. Tape cartons
  with tape machine
• Cartons are numbered and date coded
  by Marsh ink jet unit
• Conveyor to Rapistan
• Palletized 1st floor warehouse
• Shipped to Georgetown

Figure A.6.2 *continued*

A.6.6 Hazard analysis

The HACCP Team carried out a detailed hazard analysis for the entire process, considering potential hazards and their associated risk. Following risk assessment a number of hazards were considered signficant and were taken through to the next step of the HACCP Process. An example Hazard Analysis Chart for raw materials is reproduced here (Table A.6.1).

This case study is an example to illustrate the HACCP approach and is provided without any liability whatsoever in its application and use.

Table A.6.1 Hazard Analysis Chart for Mini Eggs

Process step	Hazard and source	Preventative measure
Raw Ingredients 5501 Chocolate	Metal shaving from equipment (P)*	In-line magnet located between tempering unit and in-feed line
	Sanitizer from cleaning (C)	Thorough rinse (see SOP). Training of cleaning crew
	Contaminated water leak from jacketed mild steel tank (C)	Visual observation. Preventative maintenance programme
Gum Arabic	*Salmonella* (B)* post-process contamination	Vendor certification programme/COAs
	Foreign material, i.e. filth (P)	At receiving, vacuum any filth, bags and skids as standard practice (operator training) to reject materials received in bad condition
Liquid sucrose	Yeast (B) fermentation due to dirty tanker	Raw material spec./COAs/supplier audit
Titanium dioxide/ethyl vanillin/Colours	Filth on jugs (P)	Visual observation
Speckling – tartrazine brown	Allergy	Label declare
Poly pro. film	Foreign material (P)	Visual inspection at receiving
Process steps Choc. Egg Moulding	Brass shaving and roller shaving (P)	Preventative maintenance programme
Cooling system	Mould contamination from cooling tunnel (B)	Effective cleaning. Weekly environmental swab testing
Start gumming (panning)	Foreign material: Loading conveyor (P), broken pieces of rubber	Loading conveyor is covered/operator training/visual observation
Candy coating	Foreign material unloading, uncovered trays (P)	Operator training, adherence to SOP
Bagging	Foreign material (P) when produce left on conveyor for longer time	Do not leave product on incline conveyor (see SOP)
	Loose metal (P)* bagging operation	Metal detector after bagging

P, Physical; B, biological, (pathogens and spoilage organisms); C, chemical; SOP, Standard Operating Procedure.
* Considered significant hazards after risk assessment.

A.6.7 Identification and Management of CCPs

Separate decision trees for raw material and process step hazards (Mortimore and Wallace 1994, 1997) were used in CCP identification (Table A.6.2).

Two critical control points were identified in this way. For the security of the process, an additional preventative control point for metal control on the raw material chocolate was also identified. The management procedures for these control points are detailed in Table A.6.3.

This case study is an example to illustrate the HACCP approach and is provided without any liability whatsoever in its application and use.

Table A.6.2 CCP Decision Record for Mini Eggs

(a) Raw material

Raw Material	Q1	Q2	Q3	CCP?	Comments
5501 Chocolate (P) metal	YES	YES	NO	NO	Metal detector at finished product. Additional preventative control point on chocolate
Water leak (C)	YES	YES	NO	NO	Check moisture for >2% (spec. is 1% max.)
Gum Arabic (B) (*Salmonella*)	YES	YES	YES	CCP1	Check Vendor COA (QA); long-term plan to segregate Gum Arabic preparation from 'post-cook' process steps. When implemented, this will replace vendor assurance CCP
Liquid sugar (B) yeast and mould	YES	YES	NO	NO	Process requires high temp. cooking >70 °C
All colours/titanium dioxide/ ethyl vanillin/tartrazine brown	NO				COA checked (QA)/raw material spec.

(b) Process step

Process step	Q1	Q2	Q3	Q4	Q5	CCP?	Comments
Cooling glycol brine contamination	NO						Pipe is in middle of the roller and glycol brine topped up as required by operators
Gumming/panning (P) foreign material	YES	YES	NO	YES	YES	NO	Eggs pass through sieve before bagging, non-shaped eggs fall in the bin
Candy coating/colours/speckling solution	NO						
Bagging (P) metal pieces	YES	YES	YES			CCP 2	Metal detector at the bagger

P, Physical; B, biological, C, chemical.

Table A.6.3 HACCP Control Chart for Mini Eggs

Product Name: Mini Eggs

Process step	CCP/ hazard #	Hazard description	Critical Limits	Monitoring procedures	Deviation procedures	Verification procedures	HACCP Records
Gum arabic	1	*Salmonella*	Negative	COAs and random testing every 5th load Supplier audit 2× per year	Hold, re-test DMR, send back to supplier Change supplier or ensure corrective action	Finished product testing, composite samples three times/shift	All results and records kept in QA Dept
Bagging	2	Metal	None (< 2 mm ferrous, < 2 mm non-ferrous)	MD checked every 1.5 hours	Hold, re-test Check metal detector/ incident report/DMR	Metal detector audited quarterly by QC staff/ Loma checks once a year	Records kept at BCL office
5501 Chocolate	Preventative control point	Metal	None (< 2.5 mm ferrous, < 3 mm non-ferrous)	MD checked every 1.5 hours	Hold, re-test Check metal detector/ incident report/DMR	Metal detector audited quarterly by QC staff/ Loma checks once a year	Records kept at BCL office

Approved by: _____

Date: _____

Appendix B   Pathogen profiles

The following profiles of important food poisoning organisms are reproduced with some minor adaptation from ICMSF, *Microorganisms in Foods 5* (1996). The information contained within the tables is intended as an introduction to the properties of these pathogens and should be used as a general guide only.

Appendix B Pathogen profiles

Organism	Aeromonas spp. (As)	Bacillus cereus (Bc)	Campylobacter (C)
Taxonomy	As is a gram –ve, facultative anaerobe, non-spore-forming, rod-shaped bacterium. Cells are generally straight rods with rounded ends which occur singly, in pairs and in chains. The motile As have a single polar flagellum.	Bc is a group I bacillus with a sporangium not swollen by a spore. Cell diameter is 0.9 μm.	C is a microaerophilic Gram –ve, small, vibroid or spiral shaped cell that moves rapidly, darting with a reciprocating motility.
Distribution and Importance	As are ubiquitous and can be isolated from a wide variety of sources. Generally As is considered a waterborne agent and the best sources are treated and untreated water and animals associated with water, e.g. shellfish. Seasonal variation with water temperature will cause numbers to fluctuate, As is more of a problem in fresh water than seawater. However As is also found with products not associated with seawater, e.g. raw meats, poultry, milk and fresh produce. As will grow rapidly in a chilled environment. Those most at risk from As are cancer patients and immunocompromised patients.	Bc is found in soil, cereal crops, dust, vegetation, animal hair, fresh water and sediments. The ability to form spores ensures survival through all stages of food processing except retorting. All documented cases of Bc intoxication have involved time/temperature abuse that allowed low levels to greatly increase. Normally the food vehicle is spice or cereal ingredients. The incidence of Bc poisoning is increasing.	C is harboured in the intestinal tract of wild and domesticated animals and can infect humans directly or indirectly (milk, water, meat). Raw/inadequately pasteurized milk is the most common route of infection. Also, untreated drinking water. The largest potential source of infection is poultry and is a frequent cause of sporadic illness. C has also been isolated from red meat but in smaller numbers. As C cannot grow at <30 °C, growth does not usually occur in foods. C is also very sensitive to drying, air and heat. Note: Many outbreaks have been attributed to inadequate cooking at barbeques!
Pathogenicity	As is an enteropathogen via release of a proteinaceous enterotoxin during multiplication in the intestine. Due to the virulence of As it also produces at least one haemolysin and a cytotoxin. The array of virulence factors gives As the ability to act as an opportunistic pathogen.	A diarrhoeagenic factor is a protein that can be inactivated by 56 °C for 30 minutes. It is antigenic and causes diarrhoea by intoxication after the ingestion of a large number of cells to produce the toxin. The emetic toxin is a small peptide that is not antigenic, very heat resistant (126 °C for 90 minutes), resistant to extremes in pH and enzymatic digestion.	Illness is by infection of the intestinal tract and can be caused by 500 cells in milk. Diarrhoea may be caused by C adhering to the intestinal wall and producing enterotoxin. It is a heat-labile and enterotoxin. 70% of C also produce a cytoxin which could be responsible for the bloody diarrhoea. C may invade the mucosal layer of the intestine which is helped by its rapid motility and shape.

	Organism 1		Organism 2		Organism 3	
Symptoms	*As* is associated with septicaemia, meningitis and gastroenteritis. Symptoms include diarrhoea, abdominal pain, nausea, chills, headache, dysentery-like illness and colitis. The incubation period is not known but the symptoms last for 1–7 days. Stools are watery but not mucoid/bloody. Symptoms are more severe for children than adults.		Poisoning occurs after ingestion of foods where the organism has grown and produced toxins. Two kinds of intoxication: 1. Diarrhoea 8–24 hours after ingestion of large numbers of cells/toxin, recover in 24 hours. 2. Emesis 1–6 hours after toxin ingestion, recover in 12–24 hours. Neither is life threatening to a healthy individual.		Abdominal pain, fever, diarrhoea and sometimes vomiting. Dehydration may be severe but only a problem in the young or the elderly. Incubation time is 2–7 days with illness lasting a similar time. Recurrence of symptoms is possible and *C* can be excreted for several weeks.	
Limits for growth	**Optimum**	**Range**	**Optimum**	**Range**	**Optimum**	**Range**
Temperature °C	28–35	>0–<45	30–40	4–55	42–43	32–45
pH	7.2	>4.5	6.0–7.0	4.3–8.8	6.5–7.5	4.9–9
a_w	–	–	0.995	>0.92	0.997	>0.987
Atmosphere	Aerobic	Aerobic/anaerobic	Aerobic	Aerobic/Anaerobic	Microaerophilic	
Salt %	1–2	<6	3	0.5–10.0	0.5	<1.5
Control	Moderately elevated temperatures are lethal to *As*. A low resistance to ionizing radiation is also a characteristic of *As*. Disinfectants are also effective against *As*.		Foods cooked properly and eaten hot soon after cooking are safe. Germination can be controlled by low pH, temperature and a_w. Cell multiplication during inadequate cooling of cereal/protein-based foods is a major concern. The food should either be cooled rapidly to <10 °C or kept >60 °C.		Proper chlorination of drinking water is a CCP to prevent infection as is pasteurization of milk. For poultry, care in evisceration and proper cooking of chicken should prevent infection. Cross-contamination of raw and cooked meat products should also be avoided.	

Appendix B *continued*

Organism	Clostridium botulinum (Cb)	Clostridium perfringens (Cp)	Listeria monocytogenes (Lm)
Taxonomy	Gram +ve, spore-forming anaerobic rod. Cell size 0.3 µm–0.7 µm by 3.4 µm–7.5 µm, motile with pertrichous flagella. Produces oval spores which are sub-terminal and swell the sporangium. Eight serologically distinct neutrotoxins have been identified, divided into two groups: I = proteolytic; II = non-proteolytic.	Cp is a non-motile, Gram +ve, square-ended anaerobic (microaerophilic) organism. Cp is a bacillus in Group III of the Bacillacae family. Cp can ferment lactose and forms oval central spores.	Lm is a Gram +ve, short, non-spore-forming rod, catalase +ve and facultatively anaerobic. Lm is motile at 25 °C but non-motile at 35 °C. Lm produces bluish-grey colonies.
Distribution and Importance	Ubiquitous and can appear in all foods. Spores are found in soils, shores, intestinal tracts of fish/animals, bottom deposits of lakes and coastal water. Although widely distributed the level of contamination is low or very low. However, due to the huge amounts of agricultural produce stored/processed the potential for spore growth should be fully appreciated.	Cp is found in soil, dust, vegetation, raw, dehydrated and cooked food. It is part of the flora of the intestinal tract of humans and animals. The spores of the different types vary in their resistance to heat which influences the survival and outgrowth of Cp after cooking. Outbreaks often occur when food is cooked slowly for large groups of people and then cooled inadequately. This produces the large number of vegetative cells for illness. Heat shock allows the heat-resistant spores to activate and grow after cooking.	Lm is found in soil, silage, sewage, food processing environments, raw meat and the faeces of healthy humans and animals. Infections in cattle can lead to Lm being found in milk. Wet surfaces in food facilities may harbour Lm and their ability to multiply at low temperatures allows food to be a major vector in human illness. The death rate for <60 years old is 25% while for >60 and in immunocompromised individuals it is 41%. Infection in pregnant women often results in spontaneous abortion.
Pathogenicity	Cb produces the most toxic of all natural substances. Human botulism is the result of ingesting pre-formed toxin in food.	5 types of Cp exist (A–E), enterotoxins of A and C cause acute diarrhoea in the intestine. A and C produced in the intestine cause the fluid loss due to the altered permeability of the cell membrane.	Lm is an opportunistic pathogen and is haemolytic. Three serovars account for nearly all human listeriosis. A high infectious dose of >100 viable cells is needed and infection occurs via the intestine.

	Group	Optimum	Range	Group	Optimum	Range	Optimum	Range
Symptoms	Weakness, fatigue, vertigo followed by blurred vision and progressive difficulty in swallowing and speaking. Weakening of the diaphragm is also observed and death is usually due to respiratory failure. Prompt administration of antitoxin and artificial respiration decrease the mortality rate.			Abdominal pain, nausea and acute diarrhoea 8–24 hours after ingestion of large numbers of the organism. Recovery in 24–48 hours.			A third of cases involve pregnant women and the other two-thirds affect those with an impaired immune system. Pregnant women exhibit a mild fever with slight gastroenteritis/flu symptoms with major/fatal results for the foetus. The other cases produce bacteraemia and/or meningitis with a small percentage having focal lesions.	
Limits for growth								
Temperature °C	I	30	10–48		43–47	12–50	37	–0.4–45
	II	25	3.3–45					
pH	I	7.0	4.6–9.5		7.2	5.5–9.0	7.0	4.39–9.40
	II	7.0	5.0–9.5					
a_w	I	0.99	>0.94		0.995	>0.95	0.998	>0.92
	II	0.995	>0.97					
Atmosphere	I	Obligate anaerobic			Aerobic	Microaerophilic	Aerobic	Aerobic/Anaerobic
	II	Obligate anaerobic						
Salt %	I	1.0	<10%		To 5.0, strain variation up to 8%		–	<11.9%
	II	0.5	<5%					
Control	Commercial outbreaks are due to underprocessing and/or temperature abuse and mishandling of food in food service establishments. Outbreaks usually occur after process failure or deviation from permitted tolerances. The control method depends on the type of food being processed and the Cb (A–G) being encountered. Canning, low pH, refrigeration and salt are all used to control the growth of Cb.			Most outbreaks are due to meat, therefore minimize contact of animal carcass with skin and digestive tract. Control is achieved via rapid cooling after cooking. If not sure of temperature history previously, then rapid heating to >70 °C will suffice. US regulations serve as a good guide (temp = internal): Not between 55.4 °C and 26.7 °C for more than 1.5 hours, or between 26.7 °C and 4.4 °C for more than 5 hours. For intact muscle the product must be chilled to 12.7 °C from 48 °C in less than 6 hours and chilling continue to 4.4 °C before it is packed for transportation.			Lm cannot be eliminated from the diet but the risk of foodborne listeriosis should be managed via HACCP from farm to consumer. During processing four major factors should be considered: 1. Minimize the multiplication of Lm in raw materials. 2. Destroy Lm in processing. 3. Prevent recontamination of processed food. 4. Prevent growth in foods.	

Appendix B *continued*

	Intestinally pathogenic *Escherichia coli*			
Organism	*Enterohaemorrhagic E. coli* (*E. coli* O157:H7)	*Enterotoxigenic E. coli* (*ETEC*)	*Enteroinvasive E. coli* (*EIEC*)	*Enteropathogenic E. coli* (*EPEC*)
Taxonomy	*Ec* are Gram –ve, catalase +ve, oxidase –ve, facultatively anaerobic short rods. *Ec* are subdivided into four pathogenic groups by the main mechanism causing the illness.			
Distribution and importance	There are few outbreaks of *EPEC*, *ETEC* or *EIEC* in developed countries. However *E. coli* O157:H7 has been seen an increase in the number of outbreaks in the US,UK, Asia and Canada. These outbreaks are linked mainly to undercooked ground beef and to a lesser extent raw milk, raw produce and infected fruit juice. *ETEC* is more common in developing countries and has been linked to salads with raw vegetables and to contaminated water.			
Pathogenicity	*E. coli* O157:H7 has two major virulence factors. These are two verotoxins which invade a cell and cause cell death. Colonization occurs principally in the small intestine.	*ETEC* only colonizes the small intestine and attach via antigens. Two major enterotoxins are produced which affect the microvilli of the intestine.	*EIEC* attacks the colonic mucosa and invade the epithelial cells, eventually causing ulceration of the bowel.	*EPEC* destroys the microvilli in the intestine via attachment of an outer protein membrane. The pathogenicity is also due to some other virulence factors.
Symptoms	Haemorrhagic colitis: grossly bloody diarrhoea, severe abdominal pain, vomiting, no fever. Haemolytic uraemic syndrome (HUS): prodrome of bloody diarrhoea, acute nephropathy, seizures, coma, death. Thrombotic thrombocytopenic purpura: similar to HUS but also fever and central nervous system disorder.	Watery diarrhoea, low-grade fever, abdominal cramps, malaise, nausea; in most severe form resembles cholera, with severe rice/water-like diarrhoea that leads to dehydration.	Profuse diarrhoea or dysentery, chills, fever, headache, myalgia, abdominal cramps; stools often contain mucus and streaks of blood.	Diarrhoea, nausea, abdominal pain, chills; diarrhoea is watery with prominent amounts of mucus but no blood.

Limits for growth	Optimum	Range	Optimum	Range	Optimum	Range	Optimum	Range
Temperature °C	35–40	7–46	35–40	7–46	35–40	7–46	35–40	7–46
pH	6–7	4.4–9.0	6–7	4.4–9.0	6–7	4.4–9.0	6–7	4.4–9.0
a_w	0.995	0.950	0.995	0.950	0.995	0.950	0.995	0.950
Atmosphere	Aerobic	Aerobic/Anaerobic	Aerobic	Aerobic/Anaerobic	Aerobic	Aerobic/Anaerobic	Aerobic	Aerobic/Anaerobic
Salt %	–	<9.0	–	<9.0	–	<9.0	–	<9.0

Control:

E. coli O157:H7 is found in the intestinal tract of cattle and possibly other animals. Contamination of food can therefore occur via faecal matter during slaughter or milking. Therefore control must come from proper cooking and prevention of cross-contamination of raw and cooked foods. As the infectious dose is low, the growth must be controlled via rapid cooling after slaughter/processing.

Humans are thought to be the principal reservoir and carriers for *EPEC*, *EIEC* and *ETEC* strains in human illness. They are carried in the intestinal tract of carriers and therefore infected food handlers can contaminate foods if they practise poor personal hygiene. Important controls are therefore the education and training of workers in safe food handling and personal hygiene. Also, the proper heating of food and holding foods under appropriate conditions. Plus, untreated human sewage should not be used to fertilize vegetables and unchlorinated water should not be used to clean food processing facilities.

Appendix B continued

Organism	Salmonellae (Sa)	Shigellae (Sh)	Staphylococcus aureus (Sau)
Taxonomy	Sa are Gram –ve, facultatively anaerobic, non-spore forming rod-shaped bacteria. Most are motile. Subgenus I contains the typical pathogenic Sa.	Sh are non-motile, Gram –ve, facultatively anaerobic rods which are closely related to E. coli. There are four main sub-groups which are differentiated by biological and serological characteristics.	Sau is Gram –ve, catalase +ve coci, anaerobic and exhibit facultatively anaerobic metabolism.
Distribution and importance	Those foods of animal origin and those subject to sewage pollution have been identified as pathogen vectors. Sa reside in the intestinal tract of humans and animals and colonized individuals become healthy excretors. Sa spread during the transport of live animals, during slaughter and in the processing of foods. Foods that are commonly involved in outbreaks are pork, poultry, eggs, raw milk, water and therefore shellfish and other foods that can be contaminated by animals sources.	Sh are not natural inhabitants of the environment. Sh originate from humans and higher primates and spread during the later phases of the illness via hands soiled with faeces, food or flies. Water can also become a vehicle for transmission of Sh and therefore food contaminated with water can spread Sh. Foods usually implicated are milk, salads, processed potato, cooked rice and hamburgers.	Sau is ubiquitous and found in the mucus membrane and on the skin of warm-blooded animals and 50% of humans. Sau is resistant to drying and may colonize hard-to-clean areas of processing equipment. Sau is often found in the dust of ventilation and cyclone equipment. Sau competes poorly with other bacteria and rarely cause poisoning in raw food products. Poisoning normally occurs after cooked foods are contaminated by a food handler and then held at 20 °C–40 °C for several hours. Sau is so resistant to drying that they produce enterotoxins in products with a low a_w.
Pathogenicity	Sa invade the lumen of the small intestine and multiply. To cause human illness over 125 000 Sa are needed. However, if fatty foods are ingested as few as 10 Sa are needed as the fats protect the Sa from the stomach acids.	Sh are invasive and penetrate the epithelial tissue of the intestine. The infective dose is small (10–100) to cause disease. Sh toxin is cytotoxic, entertoxic and neurotoxic. The toxin destroys the epithelial cells and promotes fluid loss. Depending on the strain causing the infection, the severity of the illness is determined.	Sau forms a wide variety of aggressins, exotoxins and enterotoxins. Sau enterotoxins have a low molecular weight and cause a vomiting/diarrhoeal response by stimulating the sympathetic nervous system. 0.1 µg/kg of toxin will cause illness in a human.

	Optimum	Range	Optimum	Range	Optimum	Range
Symptoms	Gastroenteritis, usually 12–36 hours after ingestion of the food. Symptoms are diarrhoea, nausea, abdominal pain, mild fever and chills. Possibly vomiting and headaches. Duration of 2–5 days.		Sudden abdominal cramps, diarrhoea within 1–4 days of infection. Only when the disease progresses to the colonic phase after 3 days can Sh be diagnosed. These symptoms are waves of intense cramp and frequent bowel movements producing small amounts of blood, mucus and acute pain.		Normally 2–4 hours after ingestion of food containing the enterotoxin. Symptoms include nausea, vomiting, abdominal cramps and diarrhoea. Recovery is usually in 2 days.	
Limits for growth						
Temperature °C	35–43	5.2–46.2	35–43	6.1–47.1	37	7–48
pH	7.0–7.5	4.1–9.5	5.5–7.5	4.9–9.34	6.0–7.0	4.0–10.0
a_w	0.99	0.94–>0.99	–	–	0.98	0.83–>0.99
Atmosphere	Aerobic	Aerobic/Anaerobic	Aerobic	Aerobic/Anaerobic	Aerobic	Aerobic/Anaerobic
Salt %	–	<9.4	–	<5.18	–	<21.59
Control	As low numbers can cause illness, it is important to ensure the absence of Sa from foods by: 1. A kill step in the process, especially when using raw agricultural foods of animal origin. 2. Prevent contamination. 3. Low/high temperature storage to prevent Sa growth. To destroy Sa use heat, irradiation, acidification or a combination of these factors.		Normally due to already cooked food being contaminated by infected people and then served. Handling and preparation of food is the main point of control. This can be prevented by using utensils instead of hands and good personal hygiene. Antimicrobials have limited use to control an epidemic and the risk of producing antibiotic resistant strains should be recognized.		Protect product from contamination and avoid conditions that will favour growth. If production of product involves temperatures where Sau can grow, then control of the raw materials is very important as well as strict control of the fermentation and maturation stage. The enterotoxins are made under a wide variety of conditions and are very heat resistant and survive cooking and some sterilization processes.	

Appendix B *continued*

Organism	*Vibrio parahaemolyticus* (Vp)	*Vibrio cholerae* (Vc)	*Vibrio vulnificus* (Vv)
Taxonomy	Vp is a mesophilic halophile with Gram –ve curved/straight rods (0.5–0.8×1.4–2.6 μm), actively motile due to sheathed polar flagella. Pathogenic Vp is 95% Kanagawa +ve (i.e. produces thermostable haemolysin active on human red blood cells).	Vc is an oxidase +ve, Gram –ve, often curved facultatively anaerobic rod, usually motile due to sheathed polar flagellum.	Vv is a mesophilic halophilic Gram –ve motile organism similar to Vp but will not produce lateral flagella on solid medium like Vp.
Distribution and importance	As a mesophile it is found in inshore warm coastal waters and in temperate regions during the summer. Found on all marine animals, however, most of these Vp are non-pathogenic (98%). May also appear on foods exposed to salt, e.g. salt-preserved vegetables. Most outbreaks worldwide are due to raw shell-fish and fish. Rare cases of river fish. Vp is heat labile, therefore cases due to cooked products usually due to underprocessing (rare) or cross-contamination (common).	Persist in endemic foci in Asia, especially Indian subcontinent, erupting to cause epidemics spread across world by infected humans. Spreads via contamination of raw and cooked foods and water. Vc is found in rivers and marine waters across the world. Due to sea contamination initially, seafoods are a source of contamination. Also raw vegetables contaminated with sewage by irrigation water or washing water. Bottled untreated mineral water can also be a source of infection.	Vv have been isolated from shellfish and water in the marine environment in coastal regions. Shellfish, especially oysters, are a significant dietary vehicle. All foodborne cases in susceptible individuals (chronic cirrhosis, hepatitis) are from oysters. Numbers of Vv are especially high when water temperatures >21 °C. Vv is very heat sensitive so is not reported in heat-processed foods.
Pathogenicity	Mechanisms of toxicity not fully understood. Enterotoxin implicated and may be haemolysin characteristic of Kanagawa +ve strains; haemolysin possibly cardiotoxic. Cytotoxin also involved in symptoms of illness.	Vc binds to the cells in the wall of the small intestine and produces cholera enterotoxin which causes hypersecretion of salts and water.	Causes bacteraemic and septicaemic disease, not gastroenteritis. Consists of primary septicaemia that destroys body tissue by the Vv producing heat-labile cytolysin that causes severe tissue damage.
Symptoms	Gastroenteritis with diarrhoea 98%, abdominal cramp 82%, nausea 71%, headache 42%, fever 27%, chills 24%. Onset after 4–96 hours, resolved in 3 days, low mortality.	Begins with mild diarrhoea, abdominal pain and anorexia but proceeds to copious diarrhoea and rapid loss of body fluids and body salts. Incubation 6 hours to 5 days and recovery in 1 to 6 days. With no treatment can be fatal.	Consists of fever 94%, chills 91%. In 66% of cases the severe skin lesions and necrotic ulcers develop. Mortality 40–60%. On average disease takes 38 hours (range 12 hours to several days) to develop. The course of the disease is very rapid and hard to treat.

Limits for growth	Optimum	Range	Optimum	Range	Optimum	Range
Temperature °C	37	5–43	37	10–43	37	8–43
pH	7.8–8.6	4.8–11.0	7.6	5.0–9.6	7.8	5.0–10.0
a_w	0.981	0.940–0.996	0.984	0.970–0.998	0.980	0.960–0.997
Atmosphere	Aerobic	Aerobic/Anaerobic	Aerobic	Anaerobic/aerobic	Aerobic	Facultative
Salt %	3	0.5–10.0	0.5	0.1–4.0	2.5	0.5–5.0

Control		
Prevent organism multiplication by chilling seafood to <5 °C and hold. Seafoods eaten raw should be <5 °C at ALL times. Cooking to internal temperature >65 °C will destroy Vp. Cross-contamination avoided by strict separation of raw and cooked food by preventing transfer via containers/shared surfaces/employees preparing raw and cooked products. Cooked seafood should be eaten with 2 hours of preparation or chilled <5 °C after cooking.	When organism is essentially absent it should not be found on locally produced food and GMP should provide sufficient protection, except when seafood harvested from inshore water with a temperature >25 °C, which will exhibit high numbers of Vc, although risk of disease [No. of 10^3 organisms] is unlikely. However, still best to refrain from harvesting or treat to destroy the Vc.	Due to invasive and necrotic character of the disease in susceptible individuals and the fact that it is normally contracted after consuming raw oysters, the only effective method of contol is to avoid this food. However, the risk can be decreased by suspending harvesting of oysters when the water temperature >25 °C and rapidly chilling to <5 °C post harvest. Also, irradiation treatment of 1.0–1.5 kGy to kill all Vv but not the living oyster.

Appendix B *continued*

Organism	Viruses		
	Hepatovirus (He) (Hepatitis A virus)	*Enterovirus* (En)	*Norwalk and Norwalk-like* (Nw)
Taxonomy	He and En are primarily viruses of the gastrointestinal tract. He and En are small spherical virions (22–30 nm diameter), made of a polypeptide capsid surrounding a linear single strand of DNA. He and En are resistant to acid (pH 3), detergent and lipid solvents. Hepatitis A virus is more heat stable and less acid stable than En.		Nw are resistant to acid, ether and heat (60 °C for 30 minutes). Nw have a diameter of 25–35 nm.
Distribution and importance	Bivalve molluscs are a major source of infection. Contamination can occur pre-harvest (water cress) or post-process (milk). The maximum numbers of He are in late incubation and the early clinical stages of illness. Therefore, infected food handlers can be working and are therefore a hazard. Many outbreaks are due to infected food handlers in food service establishments. Pasteurization may be ineffective to destroy hepatitis A virus.	Humans are the only important host for human En virus. En is shed in large quantities in infected persons, depending on the type of virus. The main source of contamination to food is direct/indirect faecal contamination. Bivalve molluscs retain human En due to feeding mechanism and have a much higher concentration than the surrounding environment. Due to contamination from sewage effluent. En has been found in bivalve molluscs but have not been found to transmit enteroviruses except hepatitis A. En has been found in meat, vegetables, fruit and milk. This is often due to contact with contaminated water. Contamination with an infected food handler is the other common cause. However, the route of infection is normally milk.	Transmitted by similar vehicles to hepatitis A. Viral gastroenteritis is a problem for the bivalve fishing industry. Passive transfer can affect salad, sandwiches and bakery products. Nw can be detected in faeces and vomitus in the acute phase of illness and 1 to 2 days after. Transmitted up to 2 days after the symptoms have left and is very infectious.
Pathogenicity	Hepatitis A virus replicates in the gastrointestinal tract and spreads to the liver via the blood. Damage to the liver occurs by an immunological reaction to the virus.	En infects the intestinal tract and produce lesions in other areas of the body, especially the central nervous system.	Nw invade and damage the gastrointestinal tract and produce lesions on the small intestine. Infection does not produce long-term immunity to the virus. Only a small dose is need for infection to occur.

	Aspergillus (As)	Fusarium (Fu)	Penicillium (Pe)
Symptoms	Hepatitis A has an incubation of 2–6 weeks with the onset characterized with fever, headache, malaise, fatigue, anorexia, nausea followed by vomiting and abdominal pain. Jaundice and dark urine may appear later. Incapacitation can last for several months but recovery leads to life-long immunity.		Nausea, vomiting, diarrhoea, abdominal pain. cramps plus headaches, fever and chills. An incubation period of 24 to 48 hours is experienced with an acute phase of illness lasting 2 to 4 days. Victims will feel unwell for about a week.
Control	Measures to prevent direct/indirect faecal contact are critical and rely on appropriate work practices from food handlers. Food handlers must not work when suffering gastrointestinal illness. Gloves should be worn for critical operations as hand disinfectants are not effective against En viruses, unlike soap and water. Vaccines are available and should be considered in some cases. Waterborne viruses should be kept out of food by ensuring that all water is of good microbiological quality. Shellfish-borne viral illness should be controlled by preventing harvesting from polluted water and by post-harvest treatment to destroy/remove viruses.		

Toxigenic Fungi

Organism	Aspergillus (As)	Fusarium (Fu)	Penicillium (Pe)
Taxonomy	Xerophilic and will spoil foods that only slightly exceed safe moisture limits. A. flavus and A. parasiticus produce aflatoxins.	Thirty species exist and 24 of these produce toxins which are hazardous to human health.	Fifty common species. These are classed by morphological features into four sub-genera with Penicillium spp. being most important toxigenic and food spoilage species.
Distribution and importance	A. flavus is found extensively in the environment, unlike A. parasiticus which is less widespread. Both are commonly found on nuts and oilseeds. In developed countries strict sorting reduces the risk of aflatoxins to low levels in foods. Developing countries do not have these systems in place and therefore are at risk from the aflatoxins.	Primarily plant pathogens that grow in crops before harvest only with a high a_w. Toxins are produced only after, or during, harvest. Can be a problem when crops are harvested late or left in field over the winter.	Pe with Aspergillus are the dominant fungi on decaying vegetation. Pe can grow at lower temperatures than Aspergillus and are found in the environment in temperate climates and in cool stores worldwide. Some are xerophilic but less so at low a_w than Aspergilli.
Toxins and toxicity	The aflatoxins have four effects; acute liver damage, liver cirrhosis, tumour induction and teratogenesis.	The mycotoxin is toxic due to its non-competitive inhibition of protein synthesis which varies in its method of operation according to the species. Vomitoxin, produced by Fusarium graminearum, is known to cause toxic effects in animals and human illness has been reported.	Many diverse type of toxin with a variety of molecular structures. Patulin is an example of one of the toxins produced. Two effects are produced by the toxins. Either they affect liver/kidney function or they are neurotoxins.

Appendix B *continued*

	Toxigenic Fungi					
Organism	*Aspergillus* (As)		*Fusarium* (Fu)		*Penicillium* (Pe)	
Symptoms	Exposure to acute aflatoxin poisoning produces jaundice, rapid development of ascites and hypertension. Long-term exposure to low levels of aflatoxin produces liver cancer, has a very long induction period.		Ingestion of the mycotoxin produces vomiting, diarrhoea, anorexia, tissue necrosis and damage to nerve cells. Although most show acute toxicity, some have been implicated in cancer.		Wide range of symptoms. Liver/kidney toxins cause generalized debility in human and the neurotoxins produce sustained trembling.	
Limits for Growth	Optimum	Range	Optimum	Range	Optimum	Range
Temperature °C	33	10–43	22.5–27.5	–2–35	20–24	<5–37
pH	5–8	2–>11	5.5–9.5	<2.5–>10.6	5.0–6.5	<2.2–>10
a_w	0.98–>0.99	0.65–>0.99	>0.99	0.88–>0.99	–	0.79–0.83
Salt %					>0.99	
Control	Control is focused on farm management as it has been shown that As invades the nut while in the field. Good storage practices and the final screening of the nut/crop are the final method of controlling the aflatoxin.		A lack of information exists on the physiology of the fungus or the factors which influence toxin production. Therefore, no control measures can be suggested. Vomitoxin is controlled through legislation in some countries.		Cereals/foods should be stored far from the optimum levels for growth and ideally out of the growth ranges.	

Appendix C   Glossary

This glossary has been compiled using Codex (1997) as the reference document. Certain definitions have been adapted in order to aid understanding. Where there is a significant change, the actual Codex definition is also provided.

Audit A systematic and independent examination to determine whether activities and results comply with the documented procedures; also whether these procedures are implemented effectively and are suitable to achieve the objectives.

CCP Decision Tree A logical sequence of questions to be asked for each hazard at each process step. The answers to the questions lead the HACCP Team to decisions determining which process steps are CCPs.

Cleaning in Place (CIP) The cleaning of pipework and equipment, while still fully assembled, through the circulation of cleaning chemicals.

Control Measures Factors, actions or activities which can be used to prevent, eliminate or reduce to an acceptable level a food safety hazard.

Corrective Action Action to be taken when the results of monitoring at the CCP indicates loss of control.

Critical Control Point (CCP) A step where control can be applied and where it is essential to prevent, eliminate or reduce a hazard to acceptable levels. Codex (1997) defines this as: A step at which control can be applied and is essential to prevent or eliminate a food safety hazard or reduce it to an acceptable level.

Critical Limit Criteria which separate acceptability from unacceptability and must be met for each control measure at a CCP.

Flow Diagrams Codex (1997) defines this as: A systematic representation of the sequence of steps or operations used in the production or manufacture of a particular food item.

Gantt chart A project implementation timetable. The Gantt chart shows at a glance the timing and dependencies of each project phase.

HACCP Control Chart Matrix or table detailing the control criteria (i.e. critical limits, monitoring procedures and corrective action procedures) for each CCP and preventative measure. Part of the HACCP Plan.

HACCP Plan The document which defines the procedures to be followed to assure the control of product safety for a specific process. Codex (1997) defines this as: A document prepared in accordance with the principles of HACCP to ensure control of hazards which are significant for food safety of the food chain under consideration.

HACCP Study A series of meetings and discussions between HACCP Team members in order to put together a HACCP Plan.

HACCP Team The multi-disciplinary group of people who are responsible for developing a HACCP Plan. In a small company each person may cover several disciplines.

Hazard A biological, chemical or physical property or condition of food which may cause it to be unsafe for human consumption. Codex (1997) defines this as: A biological, chemical, or physical agent in, or condition of, food with the potential to cause an adverse health effect.

Hazard Analysis Codex (1997) defines this as: The process of collecting and evaluating information on hazards and conditions leading to their presence to decide which are significant for food safety and therefore should be addressed in the HACCP Plan.

Hazard Analysis Chart A working document which can be used by the HACCP team when applying HACCP Principle 1, i.e. listing hazards and describing measures for their control.

Intrinsic factors Basic, a_w integral features of the product, due to its formulation, e.g. pH, a_w.

Monitoring The act of conducting a planned sequence of observations or measurements of control parameters to assess whether a CCP is under control.

PERT chart A diagrammatic representation of the dependency network and critical path to completion of a project plan.

Potable water Wholesome, drinkable water.

Preventative measure (see **Control Measures**)

Process Flow Diagram A detailed stepwise sequence of the operations in the process under study.

Quality Management System A structured system for the management of quality in all aspects of a company's business.

Supplier Quality Assurance (SQA) The programme of actions to ensure the safety and quality of the raw material supply. Includes preparation of and procedures to assess supplier competency, e.g. inspections, questionnaires.

Target level Control criteria which are more stringent than critical limits, and which can be used to take action and reduce the risk of a deviation.

Validation Codex (1997) defines this as: Obtaining evidence that the elements of the HACCP Plan are effective.

Verification The application of methods, procedures, tests and other evaluations in addition to monitoring, to determine compliance with the HACCP Plan.

Appendix D   Abbreviations and definitions

AEA	Action Error Analysis
ATP	Adenosine triphosphate
c	The maximum allowable number of defective sample units (two-class plan) or marginally acceptable units (three-class plan). When more than this number are found in the sample, the lot is rejected
CCP	Critical Control Point
CCFRA	Campden and Chorleywood Food Research Association
CIP	Cleaning in Place
Codex	Codex Alimentarius Commission, an FAO/WHO Organization
EC	European Community
FAO	Food and Agriculture Organization
FDA	The US Food and Drug Administration
FIFO	First in, First out – principles of stock rotation
FMEA	Failure Mode and Effect Analysis
GDP	Good Distribution Practice
GLP	Good Laboratory Practice
GMP	Good Manufacturing Practice
HACCP	Hazard Analysis Critical Control Point
HAZOP	Hazard and Operability Study
HHA	High Hygiene Area
HMSO	Her Majesty's Stationary Office
HTST	High temperature, short time
IAMFES	International Association of Milk, Food and Environmental Sanitarians

ICMSF	International Commission for Microbiological Specifications for Foods
IDF	International Dairy Federation
IFST	Institute of Food Science and Technology, London
ISLI	International Life Sciences Institute
ISO	International Organization for Standardization
m	A microbiological limit which separates good quality from defective quality (two-class) or from marginally acceptable quality (three-class). Values $\leq m$ are acceptable; values $>m$ are either marginally acceptable or unacceptable
M	A microbiological limit in a three-class sampling plan which separates marginally acceptable product from defective product. Values $>M$ are unacceptable
MAP	Modified Atmosphere Packaging
MORT	Management Oversight and Risk Tree
MRL	Maximum Residue Level
n	The number of sample units examined from a lot to satisfy the requirements of a particular sampling plan
NACMCF	National Advisory Committee for Microbiological Criteria for Foods (USA)
NASA	National Aeronautics and Space Administration (USA)
PERT	Programme Evaluation and Review Technique
PPM	Planned Preventative Maintenance
QMS	Quality Management System
RDA	Recommended Daily Allowance
SPC	Statistical Process Control
SQA	Supplier Quality Assurance
SRSV	Small round structured virus
TVC	Total Viable Count
USDA	United States Department of Agriculture
WHO	World Health Organization

Index

Lightning Source UK Ltd.
Milton Keynes UK
UKOW040045270312

189503UK00002B/5/P